T0090394

Were Albert Einstein And Charles Darwin Idiots?

A Brief Lesson In Environmental Science

Philip G. Simone

Order this book online at www.trafford.com
or email orders@trafford.com

Most Trafford titles are also available at major online book retailers.

Print information available on the last page.

ISBN: 978-1-4251-6589-5 (sc)

Trafford rev. 10/17/2019

www.trafford.com

North America & international
toll-free: 1 888 232 4444 (USA & Canada)
fax: 812 355 4082

TABLE OF CONTENTS

PROLOGUE

The way in which humans are attempting to live is unsustainable. Many of our cultural norms that are considered good, righteous and healthy are not. We have become a dangerous species, altering our physical and chemical biosphere and killing far too much life on the planet. If we learn from our mistakes and change, we will survive quite nicely. If we continue what we are doing...

INTRODUCTION

It's a new semester, my first day teaching a college environmental science course. I tell the class, pretend there is a race of aliens from somewhere distant in the universe—you know, like in *Star Trek* when Kirk discovers some weird but very highly intelligent and all powerful lifeform, compared to us earthlings. Two of the aliens are having a chat. They decide to play a cruel game: the earthlings have a choice between having half the human population randomly and immediately killed, or having all the soil bacteria and soil fungi on earth wiped out permanently. The class must make the choice for humanity in one minute, otherwise the game is off and all humans on earth will be instantly killed.

Well, most of the class goes for having all the soil bacteria and fungi killed. Only a very few students who have taken some form of biology make the logical choice, which is to immediately kill half the human species. It would be a very dangerous gamble to kill all soil bacteria and soil fungi, because if that were the case, then almost all plant life would be negatively affected to varying degrees, and a very large portion of plant life would die off.

Terrestrial plant life helps maintain climate and is the basis of all food for all land animals and some sea life. This plant life also produces approximately 50 percent of atmospheric oxygen and is the home for countless creatures, large and small. With plant life seriously damaged, most animal life, including the human animal, would die, as living systems would collapse, except maybe for the few species that could adapt. As we will see in what follows, plant life on land and in the sea in the form of phytoplankton is extremely important. Not only is phytoplankton the basis of food for life in the sea, but it also gives us much of the balance of our oxygen. It has been estimated that if we killed off all plant life, land and sea, then all air-breathing creatures, including humans, would suffocate in about 11 years (starvation and climate change would set in long before this). Considering this time frame, does it matter what the margin of error is?

There are bacteria in the soil that cause atmospheric nitrogen to be converted to forms that plants can now use in a process called nitrogen fixation. This process is vital to terrestrial plant life. Then through assimilation, nutrients are taken up by plant roots, a process that in many cases relies on a fungus living on the roots of the plant. For normal growth and development through nutrient uptake, in 75 percent of all seed-plant species and forest trees, this symbiotic, mutualistic association between roots and fungi is critical.

There are bacteria and fungi that break down dead organic matter and waste products so that many other nutrients, including more nitrogen, can be recycled back into the soil and therefore flow continuously throughout the living world. This allows plants to grow once again and makes sure excess debris doesn't pile up on forest floors. Then there are the denitrifying bacteria that return nitrogen to the atmosphere from the soil, which in effect keeps a balance of nitrogen between the two. Similar processes take place in marine environments. Bacteria and fungi perform too many functions to mention here, but these microscopic critters are very, very important. You would not want to ever mess around with what they do for us. They keep us, and them (other species), alive.

When it comes to life on earth, human beings are not the most important species. In fact, we are the least important species on the planet. The fewer people there are on earth, the cleaner the planet, and the healthier all the other species of plants and animals will be, including humans. We do not contribute anything, in net, to the sustainability of life on this planet. Quite the contrary: just about everything that we do damages life to varying degrees and therefore slowly but surely damages us. Soil bacteria are much more important than we are. So are bees, and insects overall. If all insects were destroyed, then the results would be very similar to bacteria diminishing, as most life would end. By demonstration, today's modern human being—and I emphasize modern—damages itself and all other species in order to survive. This holds true given the way we have structured our society today. Stay calm: the coming chapters will make this assessment of the modern human a bit more acceptable.

Obviously, today's environmental science didn't exist in pre-modern times. However, even centuries ago, many American Indians appreciated and in some ways understood more about their connection to the earth than many of the college students who walk into my environmental science classroom for the first time. History shows that many American Indian tribes had a great respect for the environment in which they lived. They felt a kinship with the land, plants, water and wildlife. They assigned souls and human properties to plants, animals and natural phenomena. For example, some tribes viewed the bear as Master of the Mountains, and for them, a ceremonial meal of bear was a way of bonding with the animal. Nature and its resources were revered and worshiped.

Many different Indian tribes existed and they were not perfect; they made mistakes, fought among themselves, and did impact the environment, but the American Indian generally lived in harmony with nature. Many Indians were also aware that their tribes would meet their end at the hands of the "White Man".

Today, modern humans possess enormous knowledge. Name a topic, and science and technology have explored it. For example, we know that burning fossil fuels (matter from once-living organisms) such as coal, oil, and natural gas, and fossil fuel derivatives such as gasoline, diesel, jet fuel, etc., damages or kills plants, crops, forests and water bodies, and also sickens or kills people as well as other animals. Burning these fuels is threatening life all over the planet. Global warming, acid rain, air and water pollution, and species extinction are all related to burning fossil fuels. Yet we burn more and more fossil fuel, with realistic conservation too often being at the bottom of the conversation list.

Then there are the many damaging and dangerous chemicals and waste products that we put into the biosphere, contaminating the atmosphere, soils, water, our food and ultimately all living things. More than 100,000 synthetic chemicals are on the market, from household cleaners to commercial products, and most have not been properly tested as to how they affect the human system. Every year billions of pounds of waste, much of it toxic, are released into the air, water and land. In the U.S. every year thousands of people suffer from some form of pesticide poisoning, and a small number die. In developing countries the numbers run over three million, with thousands of deaths.

We know that overpopulation of any other species damages its ecosystem and can cause the decline or failure of the species in question, as well as other species that rely on the species that diminished. At the same time, however, many of us still think that somehow humans can continue to overpopulate with no consequences. We have come to believe that we can expand and grow economies, producing and consuming ever increasing amounts of goods and services, year after year, forever. The amount of information out there about how the planet keeps us alive is overwhelming, yet we are doing more damage to its ecological systems than ever before. The American lifestyle in many ways does more damage than any other way of life, but unfortunately, other countries are catching up.

The possibility of catastrophe for humans is real, and although assigning probabilities to any one event is difficult, a growing number of scientists are sounding the alarm. Climate change brought about by global warming, and its many effects such as species extinction, sea-level rise, food and water shortages and disease—individually or collectively—could make human existence as it is very difficult. The planet is warning us as the average temperature rises and the climate changes, as the Arctic rapidly melts and starts to biologically degrade, as the Antarctic shows signs of degradation, as heat-distributing ocean currents look questionable, as glaciers all over the world melt, as northern spring and summer sea ice decrease, as sea level rises, as new diseases pop up, as species die en masse and as records are broken for temperature, greenhouse gases, hurricanes, El Nino events, tornados, forest fires and heat waves. Drought and floods are becoming extreme, and the number of people dying from all these things is becoming dramatic.

There is a chance that in a worst case scenario over the next 50 years or so, we will see frightening and detrimental changes to the way in which we are accustomed to living. I know this sounds harsh, but that's only because we've been programmed to believe that humans are invincible, that we are the ultimate in the intelligence-machine department, but are we using our intelligence in a logical manner? Shouldn't the measure of intelligence also be gauged by how a species uses the brainpower it has?

Growing up in America (and elsewhere), we have become ethnocentric as we dictate to the world's ecosystems, believing that everything revolves around the human way of life. We are taught that all is well if we have a nice place to live, money and things. Buy, shop, collect stuff, and all will be fine. As a figure of speech if you will, as long as we have a Coke, hamburger, house and pension—and our health of course—then basically life is good. When someone asks how Johnny or Mary is doing, hopefully the answer is, "Oh great! He or she has a big house, car or SUV, great job and is making good money!" Or better yet, we are very successful business people, producing and selling something in large quantities and making big money. Certainly, people like to hear that Johnny or Mary is a lawyer, doctor or teacher, some profession that is beneficial to society and emotionally satisfying. Ultimately though, it's often about the quest to get as much money as possible so that a lifestyle can be maintained and, of course, future security established. I always like to make the analogy about bumper stickers I've seen. "Shop till you drop," and, "He who dies with the most toys wins." What are these sayings indicative of? Could it be a lack of understanding of what excess consumption does to our health, and the health of our planet?

The more things we produce, consume and use, the more packaging, containers and wrappings that are thrown away, the more damage we do to the planet and its life-support systems. There is a direct physical link between how much we use or consume, and how much damage is done to life on the planet. A large percentage of environmental problems—from climate change to acid rain, air and water pollution, and human sickness and death—are represented in everything that is manufactured and used. Every single decision we make on an economic basis—to buy something, use something, and consume something—has a parallel component, an environmental consequence that we mostly ignore.

The possible self-inflicted demise of humans has been talked about for many centuries, and modern-day scientists and intellectuals can continue the discourse with a lot more knowledge to feed the discussion. I had always believed that we were in trouble as a species, but for me there simply came that day when the idea finally solidified that the human species would never survive unless its priorities were changed. We must understand that we are alive by the physical choice of the planet, and that the earth does not have to maintain that choice. The earth does not owe us a living, because it is an equal opportunity employer, and an equal opportunity destroyer. The earth throws its fits of natural disasters such as earthquakes, tornadoes and hurricanes without regard to race, color, cultural or religious beliefs, and the earth will exhibit its human-induced heat waves, droughts, floods, food shortages and sea-level rise without regard for these things again. As long as we base everything on ever-increasing production and consumption of goods and services, then we are actually attempting to violate the laws of physics and nature and cannot survive in the way that we have been up to now.

Environmental science is multidisciplinary, given that it covers areas of biology, chemistry, physics, geology, meteorology and climatology. Actually, it covers almost all branches of science to one degree or another, including areas such as sociology, history, politics and economics. Environmental science is the study of how humans affect the planet and the planet's different ecosystems, ecosystems meaning all the interacting species of plants and animals of a given area, such as a tropical rain forest. Environmental science examines the interactions of species with each other, as well as their interactions with the non-living world (non-living is abiotic, living is biotic), such as soil, air, water, sunlight, nutrients, etc. As humans we are just one single species of approximately 1.8 million known species of plants and animals, and possibly millions more species yet to be discovered, probably mostly in the world of small plants and animals, and the microscopic world of bacteria.

We are connected to the earth in the same way as other species: the same laws of nature that govern them—plants and animals—govern us. We are not so different from a dolphin, tiger, bird, elephant or any other species, other than being differentiated by our superior minds. But are our minds really superior? We've killed over 100 million of our own kind in wars over roughly the past century and still continue this practice. Over half the human world lives in poverty, and about a billion people are completely destitute and have minimal ability to obtain the most basic needs such as safe water, food and shelter. As a result over 20,000 people, 75 percent of them children, die every day from causes such as hunger and malnutrition. We are also destroying the plants and animals all around the planet that keep us alive. We are destroying ourselves directly and indirectly. We can go to the moon or build a computer, but this doesn't make us intelligent, and doesn't necessarily make our values fitting either, because if we look through nature's eyes, then all the other creatures on the planet that keep the life-support systems functioning are smarter than we are. Too many of us have become so self-centered and self-indulgent that we can't see, or refuse to see, the true picture of what keeps us all alive.

All living things are interdependent. As humans, sometimes we think that we are so advanced in every way that everything not human is beneath us. Nothing could be further from the truth: we are massively destructive and kill off so many living things on the planet that we have put ourselves in line for our own end. Look at it this way: kill off enough insect species or just a few of the most important ones, and some form of human starvation will follow. Over one-third of the food we eat is the result of plants being pollinated by insects (and we are killing many of them; more on this later). Countless numbers of plants! No, we wouldn't be capable of sending out a billion people with tweezers and eye droppers every day to do the job.

When we look at how our ecosystems work, it becomes very clear that modern humans simply will never fit in without conspicuous change. There have been many extinctions before us, and there is no reason why we can't also go that route. Throughout the earth's history many species have gone extinct in a flash of geologic time. Five great extinctions are known—possibly caused by climate change—and very large numbers, 50 to 90 percent, of species disappeared. The first was approximately 450 million years ago; the second about 350 million years ago, and the third and fourth were between 250 and 200 million years ago. The triggers for these events are still debated today. The fifth and most famous extinction crisis took place about 65 million years ago, when many species on the planet, including dinosaurs, were believed to have been wiped out by an asteroid impact. The impact blast likely started fires around the planet, but it also induced a deadly climate change by sending so much debris into the upper atmosphere that sunlight was blocked out for a long time period. There have been many climate changes throughout our earth's long history, and in recent times, even the human animal has been affected. Climate is inherently unstable yet does experience periods of stability. However, the earth's climate is always waiting for something to tweak it over the edge, and humans are doing a fine job of approaching the precipice.

An asteroid or comet impact might seem like an unavoidable event, but it is not, because today's technology could detect and possibly deflect an in-coming killer. Unfortunately, our government does not take this very real threat seriously enough either. While the U.S. government spends billions of dollars on weapons of mass destruction that could destroy all humanity (the U.S., U.K., France, Russia and China are the big arms dealers), it will not spend what is needed on systems and projects to alter the path of an incoming body that could end the existence of most life on earth. It's as though we have a tendency towards self-destruction one way or another.

Today the world is in the beginning of what will likely turn out to be a mass extinction crisis over about the next 50 years. Biology text books already recognize the beginnings of this sixth mass extinction crisis on earth. Estimates say this current human-induced extinction rate may be 1,000 times the natural background rate. (Nature has its own chaos, catastrophes and extinctions without us, making the saying "balance of nature" not completely accurate, because over time, change disturbs equilibrium. However, shorter term, stability and balance keeps some species and processes going far longer than the human life time. I will use the word "balance" in broad context throughout.) Thousands of species of plants and animals go extinct every year—possibly 20,000—directly because of human activity. The true number is not known, with estimates ranging from 5,000 to 50,000. Whatever the actual number, these extinctions are a disaster potentially culminating in our sixth mass extinction crisis. While this may not be as abrupt and obviously catastrophic as the asteroid hit that took out the dinosaurs, some scientists say this sixth mass extinction may be just as serious, with an end result that may be very similar. And it is more insidious because it is happening while we sleep in ignorance.

The earth is quite capable of supporting a certain number of human beings—carrying capacity is what it is called—and all species have a carrying capacity for the area they inhabit. We have established the carrying capacities for lots of species but so far have refused to look seriously at our own. Scientists and intellectuals have studied the earth's carrying capacity for humans, but the discussion remains a remote one, away from the public and the politicians who really don't seem to want to hear it anyway. A flowerpot can sustain only so many worms. The earth is our flowerpot and can sustain only so many humans. How many? Not the approximately 6.7 billion people (as of 2008) in their present mode of consumption and expansion habits. And certainly not the projected 2.5 to 3.5 billion more of us that are on the way over the next 50 years.

We are such a powerful force. We have changed the chemistry of the air and oceans, altered the land and inhabited or cultivated almost all of the inhabitable or cultivatable land that exists. Roughly 70 percent of the planet's surface area is water, and of what remains as land, only about half of it is able to comfortably support people. We don't live on a very large portion of the earth, yet we act as though it is an endless realm. The earth doesn't have the resources or the ability to absorb the additional damage that would come if we attempted to inhabit the rest of the land area (frozen poles, deserts, wetlands, mountains, etc.) and covered the planet completely with human societies—a ridiculous concept anyway. What little space is left on land unoccupied by us, is desperately needed for other animal species and plant species (trees, etc.), to maintain food, water, soils, atmosphere and climate. The earth doesn't have the capacity to continue much longer dealing with the human population we have right now, doing what we're doing right now.

We have seriously damaged the ozone layer and exposed ourselves and all living things to excessive deadly ultraviolet rays. We've locked ourselves into global warming. We have polluted almost everything everywhere. Yes, air and water are fresher in the Antarctic than New York City, but they're still polluted. During the 20th century the testing of nuclear weapons dropped radioactive fallout all over the world, still detected today in the atmosphere, oceans and on land. Fallout from a September 1976 Chinese test showed up in Pennsylvania. Ocean dumping of radioactive waste went on legally from 1946 to 1983 in over 50 sites in the north Atlantic and the Pacific. Thousands of tons of waste were dumped, and in 1982 alone, 435 miles off the coast of Spain, 11,000 tons of radioactive material were dumped into the sea. Although things have gotten better, the oceans still serve as the world's dumping grounds.

Burn a gallon of gasoline in New York today and within a year or so the pollution this creates will be all around the planet. As many others have said, we can't throw anything away. There is no "away." All waste comes back to us one way or another in the water we drink, the air we breathe, the food we eat, the species that die, or the climate that changes. Air pollution in the Pacific Northwest has come from China. Pollution in Canada has come from the Ohio River valley. From London to Sweden, and from the United States to Europe, and the list goes on. Pollution from here and there to everywhere. All the oceans and atmosphere are connected as one, along with the land, our earth, our biosphere.

Presently, this planet is changing. As a human support system, the earth may not continue to be so hospitable. Many of its biological systems are in decline. In a mere 150 years humans have done severe and noticeable damage to the planet's life support systems, while it took 4.6 billion years for the planet to evolve and about 3.8 billion years for life to evolve up to today. The life forms we are accustomed to took about 600 million years to evolve.

Worldwide, glaciers and ice sheets are melting, while oceans are heating up and expanding and therefore also contributing to sea-level rise. The world's oceans are exploited to the point where we are taking more marine life out of the system than nature can put back. Many fisheries have collapsed as we ruin ocean ecosystems and food chains. Marine species as well as terrestrial species are heading for an extinction crisis. Soils are being eroded at unprecedented rates. Water supplies are overused everywhere, from the Colorado River, to major aquifers in the U.S., to rivers and aquifers all over the world. Water wars may become inevitable. Climate and weather are changing through global warming, and the many associated effects are frightening. The oceans' chemistry has changed, to the detriment of marine inhabitants. Acid rain still kills countless trees, other plant life, crops, and fish, as does air pollution. The cost to human health is in the billions of dollars, without mentioning the deaths and pain and suffering from sickness related to pollution. Over half the world's original forests are gone because of human activity, and trees are still being cut and burned at a rate that keeps the net amount declining by over 13 million hectares, or about 32 million acres, per year. Therefore there is more global warming, habitat destruction and species extinction. Of what forest remains, too much of it is degraded or fragmented.

Everything we use comes from the finite earth. Look at anything around us that has been manufactured, that we bought or used, all the stuff stored in our garages, attics or basements too, everything we wear, eat, live in, work or play with or drive. Not only did all these things come from the earth, but their existence means that energy was used and some form of pollution and habitat destruction was the result. We are using up our resources and damaging planetary systems severely, and in some ways irreversibly.

We can have a bright future. Environmental problems can be solved. However, if we do not use technology on a massive scale, in time, and if we do not control population and growth while we also intimately combine environmental science into every economic decision made, we will not survive in our current form. Nature and physics will not allow it. We are producing and consuming ourselves out of modern existence.

PART ONE

1
ENERGY: WHAT DOES IT REALLY MEAN?

Everything revolves around energy. It is impossible to fully understand planetary problems and the workings of our environment without a basic concept of energy. So I'll get a little technical for a few pages, then we'll get back to the story.

What is energy? If we were in a physics class, the straightforward answer would be, practically speaking, energy is the ability to do work. The short answer, the ability to do work, gives the appearance that there is a simple answer to a simple question, but this is not the case. For the newcomer to environmental issues, a simple definition is needed, but one that is more expansive.

Energy can be measured in BTU's, a common term on new appliances such as air conditioners and refrigerators to show how much electricity will be used or how much money it will cost to run the appliance. A BTU (British Thermal Unit) is defined as the amount of heat—heat is a form of energy—that will increase the temperature of one pound of water by one degree Fahrenheit. In the world of science we like the metric system—as does most of the world—so we often measure energy with a unit called the joule. We can talk about joules of energy from burning coal, oil, gas or wood, or we could have talked about pounds of these things burned, depending on what we want to measure. For our purposes it doesn't matter how a joule is defined—crack the books if you really want to know—except as a way of keeping track, like counting money. Now, just to get a feel for energy in everyday life, let's look at food.

The food Calorie (C) that we are familiar with is also a measure of energy and is equal to 1,000 heat calories (c). There are 252 heat calories in one BTU. So if we eat something containing 100 Calories we are getting the energy equivalent of about 400 BTUs or 100,000 heat calories, which is the amount of heat that would be released if we literally burned that 100 Calories of food. This is what your body will do, burn the food chemically to gain energy. Considering there are about 4.186 joules to a heat calorie, the 100 Calorie portion of food eaten contains 418,600 joules of energy to use up within the body. A gallon of gasoline contains approximately 132 million joules of useful energy. There is a lot of energy in gasoline; that's why a car runs the way it does—but please, don't drink gasoline when you are feeling low.

The energy that we want to use to produce electricity is generally in the form of heat: to boil water, produce steam, and spin a turbine and generator. However, the initial energy forms don't always start out in the form of heat. Nuclear energy generally relates to uranium. Split the uranium atoms (fission) and useful heat energy is released along with radiation. Wood, coal, natural gas, oil and oil products such as gasoline, jet fuel and heating oil hold chemical potential energy. Burn this stuff and useful heat energy is released. The law of conservation of energy reminds us that energy can't be created or destroyed; it only changes forms. The law of conservation of matter reminds us that matter can't be created or destroyed, but it too only changes form. Make electricity or drive a car and the energy stored in some form of matter—gasoline in the car for example—is transformed from one form of energy to another, not destroyed.

Okay, so we burn some coal. It turns to air pollution, ash, soot, etc., and some energy, and of course, once we use up the coal, it's gone, and its available stored energy is used and released, never to be used again. The energy has changed from the chemical potential energy stored in the coal before it was burned, to heat energy after it was burned.

There is a fixed amount of energy in any closed system. The universe might be one such closed system. The earth is a closed system—ignoring the sun for a moment and the earth's internal energy—and so is the room you are in if you select the walls, ceiling and floor as boundaries. The amount of matter in any closed system dictates how much energy we will have available to make electricity, power planes, trains and cars, etc., or to manufacture everything we need and want. This concept gets a bit complicated if we now factor in the sun, but matter is still involved because the sun's energy is the result of hydrogen being fused into helium, 93 million miles away. Matter and energy are still being used and transformed as the process sends radiant energy to the earth.

Energy cannot be conserved, and at first this statement does seem strange. This is because energy is already conserved by the fact that there is a fixed amount of matter from which we can get it, and therefore a fixed amount of energy resulting from that matter. Even if we are very careful about how we use energy—in other words, very efficient—the quantity of matter and therefore the amount of energy available to us does not change; we can only make that energy go further. Even sunlight is not limitless energy because the giant fusion furnace using hydrogen at a rate of about 600 million tons per second will run out in about five billion years. This is why solar power is such a beautiful solution, as will be discussed more in chapter 4.

For humans, most of the energy that we choose to use is what is available to us, stored in the paper thin crust of the earth in the form of fossil fuels, uranium, wood, etc. Keep in mind that all other vital things we need, such as metals, concrete and synthetics, all the things around you, also come from the crust.

The earth is approximately 7,900 miles in diameter, yet most of the resources we obtain from it come from only the top one or two miles, sometimes a bit more, as with deep oil wells and mining operations. Soils to grow food, trees and all plant life are contained in the top few feet, and ground water levels can be measured in hundreds of feet, although it can be deeper. The breathable atmosphere is also very minimal, only about two miles thick. Above this, not too many people regularly roam. The easily breathable atmosphere where one does not have to seriously acclimate to higher altitudes, is about a mile or so thick. Relative to the diameter of the earth, it too is paper thin. Our atmosphere is not as extensive as some may think and is very easy to alter.

If we drew an eight inch diameter circle on a piece of paper representing the earth to scale, the relevant thickness of the breathable atmosphere or the usable crust drawn to scale at the surface of the eight inch earth would have to be such a thin line that it would be difficult to see. So, other than the upper crust, the rest of the earth's interior has nothing for us, although it does have value as a producer of our magnetic field, without which we might lose our atmosphere, and of course gravity is nice too. We also need volcanoes and plate tectonics for deeper reasons (pun intended), but these are areas not for this book.

So, we turn matter into energy, then this energy takes a changing path. I'll use very unrealistic numbers, but this does not change the significance of the concept. Suppose the sun sends down 10 joules of radiant energy and as a result a plant grows. Chemical potential energy is now stored in the plant matter through the process of photosynthesis. The plant actually turns the sun's radiant energy into a different form, that is, chemical potential energy. It becomes obvious when we look at a plant that it has energy because the plant could be burned to release energy to do work, or we could eat the plant to gain energy to do biological and physical work. Photosynthesis is an energy storing process by which inorganic materials are turned into organic materials. This process defines plants as they take in nutrients, carbon dioxide and water—and the sun's radiant energy—to produce oxygen and food that humans and other animals use, along with other vital services such as soil development and preservation, water recharge to underground aquifers, maintaining climate, as well as housing the other animals of the world.

Now, our hypothetical 10 joule plant dies and through complex chemical and geological processes over a very long time changes into coal. We mine the coal and then burn it to get heat energy to do work for us, usually by producing electricity. Once the coal is gone, it's lights out. Thank you to the sun and the original plant and chunk of coal, but we can't wait around for 100 million years or so for more coal to form.

Okay, here comes the unfortunate truth about our little energy scenario. We started with 10 joules of radiant energy from the sun, but we do not end up with 10 joules of energy to do work, as in toasting toast or turning on a light or computer. If we did, we would be violating the first and second law of thermodynamics. The first law of thermodynamics is really a repeat of the law of conservation of energy (energy can't be created or destroyed) as it applies to heat and work. The second law of thermodynamics ultimately says that in every energy conversion—that is when energy changes from one form to another—some useful energy is always unavailable. There is absolutely no way around these physical laws of the universe.

There's a little thing called friction that uses up energy along the way, through all energy conversion processes. I'm not only talking about friction in the physical sense, as when two moving mechanical parts of a machine rub against each other. There are other ways in which energy is lost, such as simple heat flow from a hot item to a colder one, or in the biological process, when we eat the plant and turn it into energy for growth and work. The point is, to gain useful energy from any process, some of the original energy must be used up so that useful energy is left over to do work.

So, the sun sends down 10 joules of radiant energy, but the plant uses some of this energy in order to grow and maintain itself through the process of respiration, which is an energy releasing process. You and I do the same thing; we respire, that is, breathe in oxygen and burn food to grab its energy. Again, for simplicity, let's say the plant's process of growing used up four joules of energy (the true amount varies), so now we have approximately six joules stored as chemical potential energy in the plant. Eventually the plant dies and luckily decides to turn into coal. The coal forming process now uses up some of this stored energy, which we will randomly call one joule, so our coal now stores five joules of chemical potential energy, from the original 10 joules sent via the sun.

Now we burn the coal and turn some of it into heat energy in order to boil water and create steam, but heat is lost to the surrounding atmosphere through the metal parts of the boiling device and through steam transport. If we are very careful, four joules is captured as useful energy. The steam produced spins a turbine, so the heat energy has transformed into kinetic energy—the energy of motion—but the friction of the moving parts of the turbine, like bearings and axles, cause another half of a joule of energy to be lost as heat. We are down to 3.5 joules of kinetic energy left to do work for us. The turbine turns the generator, which also faces some friction related to moving parts, and therefore more frictional heat energy is lost, say another half a joule, and finally electricity is produced. Three joules of electrical energy is flying through wires to our home, but there is still no free lunch. As electricity moves through a wire, resistance on a molecular level is encountered so that some of the electrical energy is lost along the way as it is converted to heat energy; that's why wires get hot. Now the electrical energy hits our toaster, and if we are lucky, we have 2.5 joules left to toast our morning bread and when that's done, the last of the energy is finally lost forever as heat to the universe, or environment, whichever you prefer. Energy was not created or destroyed. We started with 10 joules and ended with 10 joules as heat loss, the issue being we only got to use 2.5 joules of the original energy to do work—that is, toast our bread, cook, use power tools or do anything else electrical. This scenario is similar for all energy conversions. I repeat: all energy conversions. What we've just seen is classic physics, that is, different forms of the exact same thing—energy! Whether it's in the form of the sun's radiant energy or some other form, such as chemical, kinetic, heat or electrical, doesn't matter. It's just a matter of how we use it, and inevitably lose it.

Primary energy comes from a process, such as burning coal, oil, wood or fissioning uranium. Secondary energy can be electricity, the energy we get from the primary process. In general, it takes three units of primary energy to generate one unit of secondary energy when producing electricity, and of course this will vary depending on what technology we use and how efficient it is. Nevertheless, that three to one ratio is a sobering look at how much energy must be lost during energy conversions—as much as 60 to 70 percent—to get some useful energy to us, the end users. So when we see light filling the room, that radiant energy might be only about 30 percent of the original energy we started with, usually from fossil fuel burned.

Think about recycling. It saves additional resources from being extracted from the earth, but recycling still requires the use of energy and resources such as coal, oil, water and chemicals. It is a manufacturing process. Although less energy might be used in recycling the original product than in manufacturing it from scratch—in some cases substantially less energy—energy is still used to re-melt metals, or to clean and recycle cans, plastics, glass and newspaper where inks must be removed. Many things are recycled, from tires to batteries, cars, airplanes, electronics, demolition debris, and industrial and manufacturing waste. Conveyor belts are run, machines are used, lights are on and trucks are driven, all operating on something, usually electricity that comes from burning coal, and diesel fuel to run the truck. So pollution and waste are still released into the air, water and soils as things are processed, packaged and transported.

In the U.S. over 20 billion plastic drink bottles are sold every year, and some are recycled; however, many of them hit the garbage as people are on the move. In 2006, the U.S. recycled about 82 million tons of materials, all of it with an environmental cost. Recycling is important and very necessary as it slows the filling of landfills and incinerators and reduces some pollution, but it is not the panacea. It will not allow us to grow forever.

We can't win, tie, or leave the game. I did not invent the ballgame analogy, but here's the situation. In the energy game we can't win, meaning that we can't get more energy out of a machine or any system than we put in. Machines just make things easier for us to do; they do not create energy or save any work. You can't tie, meaning we can't even get the same amount of energy out of a machine or any system than we put in. We always get less. So we always lose the game! We can't leave the game because we need energy to live, so we have to keep going after it, and we have to keep using energy to get energy. This is why efficiency is so very important to everything that we do and all machines that we use. In other words, reduce friction and waste, and find ways to get as much useful energy out of all the matter we use.

We require energy not just to drive, turn on the lights or to manufacture everything we need and want, but also physically to live. Sometimes the energy used to obtain food is more than the energy gained from eating food. The amount of energy used to produce food, package, ship and refrigerate it must also be factored in when figuring how much energy was really used to get us to walk up the flight of stairs, to run around the block, or to write a book. If we now eat the food that was derived from the plant to which the sun originally gave 10 joules of energy, we find that energy conversions in our bodies are similar to the coal scenario. Where we stand in the energy chain depends on what we eat. The most efficient way to eat would be to eat an easily digestible plant with no processing between the ground and plant, and us, so that energy waste is minimized.

Let's say we eat a carrot from the garden. In reality there was some energy loss as we burned energy to pick it, clean it, and lift it to our mouth, but we can forget that for practical purposes because the energy used is negligible. Now our body processes the food through respiration, the energy releasing process (some energy may be stored for later use). The energy stored in the carrot through photosynthesis is now released so that we can do work: run, play, do our jobs, grow, and operate all our bodily functions. During respiration, we inhale oxygen and oxidize or burn food, similar to burning coal to release energy—minus the fire of course. Except our body burns the food in small packets as we need it, and we breathe out the by-products carbon dioxide and water vapor. The heat energy we feel coming out of our body is gone, according to the second law of thermodynamics. At this point we should call this short energy chain a food chain. Lengthen the food chain and the energy loss increases, so the number of people we can feed with the same carrot reduces dramatically.

The base of the food chain, the energy supermarket, is all the stuff that grows, basically the green plants, whether we eat them directly or eat an animal that first ate the plants, or eat another animal that ate the first animal that ate the plants. The base of the food chain is ultimately all the things that get their energy from the sun, to pass on to everything else. The base of the food chain in the ocean is phytoplankton—many different types and sizes—the floating plant life that also photosynthesizes. Larger zooplankton—small floating or swimming organisms—eat the phytoplankton, and then a small fish eats the zooplankton; a larger fish eats the smaller fish and maybe the process goes one more level up in the marine food chain to a larger fish, a top predator such as a shark, and that's it. Some marine life, such as particular whales, eat massive quantities of the zooplankton directly. Energy and food chains are extremely complex, and there are volumes written on the subject.

Back on land, terrestrially speaking, if we first fed our carrot to a rabbit, and then a fox ate the rabbit, then by the time we ate the fox the energy loss from the original carrot would be enormous. Remember that all the original energy came from the sun but now we make the food chain longer, so from feeding level to feeding level (called trophic levels) more energy was lost, in many cases 90 percent from level to level.

So let's say—again, I'll use an unrealistic number, I like 10—the carrot has 10 joules of energy stored in it. The rabbit eats the carrot and through the respiration process 90 percent of the energy is lost and so the rabbit has only 1 stored joule of energy from the carrot's original 10 to pass on. The fox eats the rabbit. Now the fox respires too, and so eventually has only one-tenth of a joule stored from the original 10 joules of energy to pass on to the human who eats the fox. This is why, in nature, food chains are always short, like from grass to a deer to a bear (yes, a bear will eat a deer). Or from tree greens to an elephant, or from small fish to medium fish to large fish. Food chains are necessarily short because as we go higher up the chain the useful energy available becomes very minimal. This is also why an animal must eat a lot of food to gain energy, and why the plant biomass, the quantity of green stuff, must always be much greater than animal biomass. And indeed plant biomass represents about 98 percent of our planetary biomass. Every feeding level will always have more biomass than the feeding level above, even from animal level to animal level (exception: phytoplankton can temporarily have less biomass than the zooplankton that eats them, because phytoplankton reproduces so fast), otherwise there wouldn't be enough energy to pass on up the food chain. Put simply, one or two large fish that are on the top of the food chain might rely for their survival on a billion microscopic producers (plankton) at the bottom of the food chain.

Let's look at something more identifiable. About 45 percent of the grain in the world is grown as animal feed. It takes about 1,000 tons of water to produce one ton of grain, and 70 percent of fresh water used by humans goes to irrigating crops. If we fed people more grain products instead of meat, water would still be used, but the tremendous pollution and excess water and energy use from raising and manufacturing meat would stop. Worldwide there are more than 25 billion animals for human consumption, many of them chickens. About 1.5 billion are cattle and buffalo. Since 1950 world meat production has increased five fold. Many experts believe, worldwide, cultivating land for food production in general may have reached or exceeded its sustainable limit.

In the United States 70 percent of the grain crop produces feed for animals. This satisfies our growing population's addiction to meat, especially red meat such as hamburger and steak, and also dairy. There are approximately 100 million head of cattle in the United States, about one cow for every three people, and most of them are raised in massive feedlot operations. The energy loss from corn, for example, to cow as a result of the cow's metabolic functioning is, you guessed it, approximately 90 percent. That is, of every 100 food calories in corn (or other grains) that the cow consumes, only about 10 are left in the cow to pass on to humans. We could feed many more human beings—assuming humane economic distribution of the food—and dramatically slow world hunger and starvation if we stopped eating red meat. (The efficiencies in meat production as far as grain used to obtain animal weight, vary, and from this standpoint, chicken is a much more sensible choice.) If we just cut way back on the amount of meat that we eat, we could still feed a greater number of people with the grain that was otherwise fed to the cow—and we would be much healthier as a heavy red meat diet can be deadly to the heart, among other things.

Corn doesn't have the protein that humans need, so instead of growing so much corn, we could increase the amount of beans—soybeans for example—that we already grow. They contain the protein without the fat that comes from red meat.

In any event, the meat diet as it exists today must eventually diminish, because as world population and consumption continues to grow—and especially if climate change messes with agricultural yields, as it likely will—we will not be able to feed so many people with such an expensive, polluting process that wastes space and food calorie value. Additionally, this type of meat diet is a practice that only richer countries can afford, and where the environmental damage from this is generally ignored. But in some poor countries where the food chain is necessarily short and a vegetarian diet is the order of the day, both the monetary and environmental costs of factory farming meat are difficult to ignore. However, as many developing nations industrialize, among those who can afford it the meat diet has rapidly caught on and the problems that the processing of meat causes are becoming evident. In China cholesterol levels that were normally as low as 150 have risen in some areas dramatically. With over 1.3 billion people, China now leads the world in meat production and consumption. However, meat consumption per capita (per person) in developed nations similar to the U.S. is still higher than in developing nations.

The U.S., as well as other countries, has encouraged meat consumption all over the world. The meat industry has become global. A cow is an oil burning machine. It has been estimated that in the United States, every steer raised for meat requires approximately 300 gallons of oil in its lifetime. This oil use comes from transportation, oil-based fertilizers, herbicides and insecticides (pesticides) to grow the feedlot grains, as well as various machines and equipment used. We must also factor in electricity use, so more fossil fuel (or nuclear) is likely used.

Then there is the waste. In the U.S. alone approximately 1.3 billion tons of animal manure is produced each year. Much of this is a source of water pollution from manure run-off into thousands of miles of rivers and streams and into groundwater. The Environmental Protection Agency (EPA) asserts that this is a widespread source of river and general water pollution in the U.S.—and people do get sick from it. This pollution also contributes to oxygen depletion of waterways, fish die-offs and algae blooms. Also, annually about 100 million tons of methane (CH_4) is released from cows' belching and flatulence (go-ahead and laugh) and another 30 million tons/year of methane is released from the anaerobic (bacterial breakdown without oxygen) decomposition of manure. Methane is a major greenhouse gas, more powerful than carbon dioxide. As hard as it is to believe, the large number of domesticated animals does seriously add to global warming because of the vast amounts of methane they release.

In Latin America alone over 50 million acres of tropical rain forests have been converted to cattle pasture. In the tropics in general, massive deforestation releases annually to the atmosphere, 1.6 billion tons of carbon which equals about 4.8 billion tons of our favorite greenhouse gas, carbon dioxide. In Brazil, increasing overseas beef sales from 1990 to 2000 have caused destruction of the Amazon rain forest to increase from about 102 million acres to about 145 million acres.

Ultimately, the cow we raise for meat wastes a lot of energy and resources. The meat production process burns fossil fuels, pollutes the air, water and even soil, takes food away from other people, and can clog our arteries and mess with our body chemistry. I'm not saying be a vegetarian, but eating large amounts of meat is very unhealthy for us and the planet. Energy must be passed on among all living things. The issue is not that we pass energy through a food chain but rather how we are doing this, and in what quantities, due to human overpopulation and overconsumption.

Okay, energy can't be conserved, and we waste enormous amounts of it. That's why the key to the ball game for any type of energy use is efficiency, or again, how much we reduce friction and how carefully we use energy. Drive a high mileage vehicle, walk once in a while, shut off the lights and computer, lower the air conditioner, drive slower, forget all the unnecessary gadgets like the Jet Ski, off-road vehicle, electric knives, can openers, hair blowers etc., lower the thermostat, use a rake and broom instead of a gas or electric blower. Use energy efficient technology across the board. Well, you get the picture.

Think of the gasoline in our cars. Thirty miles to the gallon sounds good? Not when 60 is available in a hybrid as it is now, and not when we finally realize how dangerous it is to burn gasoline. The typical internal combustion engine in a car or truck is not very good at converting matter to energy, though the rate of conversion varies from engine to engine. Even Exxon-Mobil realizes that a typical car is on average only about 20 percent efficient—trucks are inefficient too. The number can easily be lower, but at 20 percent, this means that 80 percent of the useful chemical potential energy stored in gasoline is lost, much of it as heat energy from the time the gas is burned in the cylinders until the time some energy gets to the wheels. This lost heat is what we feel radiating off the engine and wheels and coming out the tail pipe, but the loss is actually much worse. How much energy was used to pump the oil out of the ground, ship it, refine it into gasoline, truck it to the gas station, pump it into the underground storage tank and then pump it out of the same tank into our car's tank? The answer is, a lot! The whole process burned coal for electricity as well as gas and/or diesel fuel that was refined from oil. Wait, we have to throw in the people who drove to work at the oil fields and refineries and burned gas in their own cars. The true efficiency of a 30 mpg vehicle is far less than 20 percent and consequently far less than 30 mpg. Driving large, wasteful vehicles is not reasonable. Using any device that isn't as energy efficient as possible is not rational. And buying bigger items, such as vehicles, houses or any product also causes the use of much more energy in the manufacturing process. Then there is the maintenance of the bigger product, like a vehicle. There is more oil in the crankcase, more water to wash it, more stuff to throw away or recycle at the end of its life cycle.

Everything that is manufactured, from a pencil to clothing, food, plastics, metal, cars, planes, chemicals, building materials, boats, toys and on and on, is manufactured using energy and matter and so ultimately pollutes ground, water or air in one way or another. This pollution can be emitted indirectly from burning coal, natural gas or oil—nuclear fuel, too—to generate electricity to run manufacturing systems, and it can be emitted directly from manufacturing machines running on fuels such as diesel or fuel oil.

Even running water to wash our hands or take a shower uses energy and so pollutes the air and also adds to global warming, because the water is being sent to us with some kind of motorized pump that is generally getting its energy from fossil fuel. Then there are the facilities where people work with lights and computers, and where labs are used to analyze water, and where chemicals such as bleach are added to the water, some of which evaporates into the atmosphere. Water that is disposed of down the drain is pumped to, and processed in, a treatment plant—the same place most wastewater goes at least in developed countries like the U.S.—and then sometimes pumped out to the ocean, all of which means more electricity, motors, fossil fuels and added chemicals such as bleach in our environment.

The fecal matter that goes down the toilet bowl gets taken out of the waste stream and processed into sludge at the treatment plant. Thousands of trucks run this waste to landfills every day all over the country, adding to pollution and habitat destruction as more and more landfills have to be created, not just for sludge but for the vast amounts of garbage in general that our society generates.

Plastics not only use oil as an energy source to be manufactured and to be shipped, but plastic for the most part is chemically produced from oil in the petrochemical industry. Oil is an organic compound. PVC or polyvinyl-chloride, a very common plastic produced petrochemically is called a synthetic organic. There are thousands of synthetics produced by the petrochemical industry, from plastics to chemicals, fertilizers, household cleaners, paints, cosmetics, pesticides, medicines, clothing and fibers. These synthetics and almost all materials we use—from metal to wood to concrete—rely on crude oil directly or indirectly. Ultimately, oil is related to almost everything we use, manufacture or transport. This is why the price of so many different material items goes up when the price of oil goes up. The more stuff we use and the more we use things inefficiently, the more oil—all fossil fuels—we use and the more environmental damage we do through chemical and fossil fuel pollution at the manufacturing sites and from product use and disposal.

Buy a paperclip and use it, and you just damaged the environment. Stare at that tiny paperclip and come to understand the energy chain and waste chain represented by the piece of metal. Appreciate the bulldozing, mining, processing and habitat destruction that is part of all manufacturing and use of materials. Look at the packaging that is constantly thrown away, and think about the damage that this does. I am not saying that human beings should live in caves again and do nothing. There are many ways to live in balance with nature. I am simply trying to bring out the ramifications of being a modern human, which in many ways makes us unnatural to the earth, while paradoxically we are still attached to and dependant on the earth.

The United States per capita resource use is excessive, and the U.S. still uses more oil than China, but this may soon change. China—with about 1.3 billion people—will soon surpass the U.S. in total energy use. Not because China's population is as wasteful as ours—although they seem to be heading in that direction—but because their larger population is becoming massively industrialized. India too, with over 1.1 billion people, is on the road of heavy development and huge energy use. The environmental damage China, India and other developing nations are doing is severe. In the near future this damage will be even more swift and catastrophic than what has already taken place, in part because in order to develop, these countries don't have to take the time to create technology along the way, as did the United States and Western Europe. Machines and manufacturing techniques are already on the table for the taking and improving. The progress of undeveloped countries is becoming frighteningly and dangerously rapid, and so far they are still basing most of this growth and expansion on fossil fuels.

Countries such as China have started the process of understanding the environment. Using solar panels, wind power and better building insulation and allowing easier aquifer recharge through porous pavement materials, along with their attempt at population control, are some of the things China is exploring. Change is starting to take place. In spite of this, it should be noted that China is nowhere near saving the environment—they are seriously damaging it. As of 2007, 90 percent of China's energy, and about 70 percent of India's energy, comes from coal burning, and this is expected to continue for a long time, possibly 30 or more years unless they radically and rapidly change. If China and India were to catch up to the U.S. in current per capita resource use, which appears to be their goal, it has been estimated that we would need another planet earth to support the two economies.

Old news: the United States, with about 4.5 percent of the world's population (about 303 million people out of 6.7 billion), is responsible for approximately 25 percent of the world's energy use and produces about 25 percent of the world's greenhouse gases. In 2005 the world burned about 5 billion tons of coal while the U.S. was responsible for burning about 20 percent of this. The technology exists to change this trend, but so far the United States has chosen to change little. One American uses more energy and does more environmental damage than do 20 or more Africans. However, in Africa and other similarly poor areas other types of damage occur, not so much from modern consumption habits but instead because of too many people living in poverty. For example, most of the deforestation in the world takes place in developing countries, to make room for agriculture or pasture, to provide wood for export, and to obtain wood to be burned for cooking and heating in poor rural areas. Plus, too many people hunting is devastating animals and habitat in Africa; the bush meat trade will be discussed later.

In the U.S. there are too many lights, appliances, SUV's, overpowered cars and oversized houses, in addition to stuff that we buy, use a few times and throw in the garbage or garage to eventually be trucked away as waste. We have too many inefficient and wasteful ways: we leave lights, computers, televisions and air conditioners on and use unnecessary gadgets and toys, as well as devices with standby power that constantly draw unnecessary electricity. Go into any department store, sporting goods store—even a food store to some degree—or most any store for that matter and look around. It's a safe bet that half the stuff on the shelves is completely unnecessary, but we want things that make us feel good, things that drive the growing economy. This constant consumption all adds up to huge energy use, pollution and habitat destruction. If the rest of the world used resources and energy like the United States, the world's life support systems would collapse, and yet much of the world is trying its best to mimic us economically and pick up our consumption habits.

In many ways the U.S. is the most environmentally damaging country in the world, not only because of our lifestyle but also because of the mentality of excess consumption that we export, and have always exported on a psychological level as a status symbol to be desired and copied. Instead of teaching the world to conserve, we push everyone to be as wasteful as we are, while at the same time U.S. energy needs alone will increase by about 60 percent in the next 20 to 30 years if we keep growing and expanding our population and economy at today's rate.

China, with four times the U.S. population and its own industrial revolution, is continually striving to boost its gross domestic product (GDP) and so is exploding its economy. GDP is the annual value of all goods and services produced within a country. As of 2005 China has surpassed the U.S. in yearly coal burned and has surpassed the U.S. in many commodities used, such as steel and grains. They are now producing more carbon dioxide than the U.S., and their contribution to species extinction, deforestation and air and water pollution is severe and dangerously growing. Many areas of China are extremely polluted, and sicknesses from air and water pollution are common.

The other bad news is at the current rates of growth, world energy use will increase by about 70 percent by the year 2030, and most of this use will come from fossil fuels if we stay on our current track. The fact is that coal use has increased every year for centuries. Furthermore, world energy demand will only increase as time goes on, and may actually increase by 400 percent or more over the next 50 to 100 years, assuming humans remain completely viable for that long. Even allowing for a margin of error, these types of projections continually remind us that our approach to this planet and our societies has to radically change or something will soon go bust.

In the U.S. about 85 percent of our total energy use comes from the fossil fuels crude oil, coal and natural gas. If we take only electrical energy into account, still the majority of it in the U.S. is generated from fossil fuels. Coal burning generates a bit over 50 percent of our electricity, and burning natural gas generates about 17 percent of the total, while burning oil produces about 3 percent. Roughly 20 percent comes from nuclear and about 7 percent is from hydroelectric power (dams). Renewables (bio-mass, wind, solar, etc.) produce the balance of

U.S. electricity, unfortunately a very small percentage. In 2005 the U.S. burned about 1 billion tons of coal, more than 90 percent of it for electricity. Oil is both the raw material and the energy source for countless manufacturing and chemical processes with its most obvious use being transportation in cars, trucks, aviation, and shipping (oceangoing vessels are extremely polluting) and home heating.

Most of us have been programmed not to think about energy, but matter and energy are not limitless, and the earth's ability to absorb waste and pollution from constant and increasing energy use isn't limitless either. Keep something else in mind: everything around us—every house, apartment building or commercial building—will someday have to be torn down and rebuilt. Cars, trucks, boats, toys, clothing and recreational items, bridges, roads, everything will have to be junked or disposed of, and most of this will be manufactured again. Nothing lasts forever. Therefore, an additional quantity of energy and matter—pollution and environmental damage—is needed to manufacture, build, maintain, use and dispose of all our things, even without factoring in any further economic growth.

It is often said that we will inevitably become more efficient—this is true—and use energy more wisely with advancements such as higher mileage cars and more efficient lights and appliances, and that this is the answer to our problems. But the added efficiency does not matter if we double the population while we simultaneously keep increasing the number of consumers and the quantities of goods they consume. If those increases continue, we will keep using increasing amounts of energy and resources even if the incremental demand per person actually does go down. All we are doing is delaying the inevitable. There are not enough resources on the planet to support the growth that is taking place.

A stunning example of the massive waste that is typical of the U.S. throwaway society (other countries too)—called an economy of waste long before this book—is junk mail. Everyday in the mail we all receive envelopes, free newspapers, magazines and postcards that are loaded with advertisements. First, we cut down over 1.5 million trees every day for paper products, using devices that run on some kind of oil derivative, gasoline in the chain saw or diesel fuel in tractor-like machines. (About 4 billion trees/year worldwide are cut.) Don't forget that people drive to work here, too. So overall there is lots of air pollution, water pollution, habitat destruction and contribution to global warming and species extinction. Even when they are not driven to extinction, species don't always fully come back when we fragment their habitat with roads. Even if trees are replanted species are still disrupted because their habitat has been changed to something they are not accustomed to, that is, different tree species, monoculture, different tree sizes, and a constant alteration of animal's habitats and food supply.

Next we ship the logs by truck and therefore cause more pollution. Then we process the logs in a factory by cutting, pulping, bleaching and processing the result into paper. More electricity is used, so more fossil fuel is likely burned, while lots of water and chemicals are used. We now have air and water pollution—and don't forget that chemicals are spilled, disposed of and evaporate, all of which adds to pollution.

We now have paper that is shipped to another manufacturing plant, so more air pollution is generated in the transportation process. Then more pollution from more processing takes place as we cut the paper and use more manufactured chemicals in the form of inks and dyes to print advertising on the finished product. Maybe a free pen or some other giveaway is added, and don't forget the plastic window on many of the envelopes.

Now the junk mail hits the post office, so—more pollution—jets are flown and trucks are driven all over the country every day to deliver hundreds of millions of these envelopes, newspapers and magazines to every house, apartment and commercial enterprise. Then the consumer goes to his or her mailbox, takes these items and ultimately puts all of them straight into the garbage, mostly without even looking at them. Now a garbage truck, which usually runs on diesel fuel—more pollution—picks up this garbage and drives it to a landfill, or sometimes an incinerator, where the paper products are dumped or burned. Even if some of this junk mail is recycled back into usable paper, the overall pollution and energy use from the recycling is considerable. Ultimately, eventually, it all ends up in a garbage truck.

A much better, far less polluting and wasteful option would be to cut down the tens of millions of trees every year, bypass the whole process of junk mail, and just bury the trees directly into the ground as garbage; at least they would decay and return the nutrients that they originally took out of the ground. Ridiculous, of course, but far more efficient and less polluting than processing trees into junk mail. Obviously, skipping the process in the first place makes the most sense. The whole scenario of trees to junk mail is as irrational as a society can get, and it is representative of a society that is generally irrational and irresponsible in many of its consumption habits. To allow such pure waste and pollution, which adds to the sickness or death of plants, people and other animals through all the various effects of environmental degradation, including global warming, is thoughtless. We have a culture that too often says if doing something makes money then it must be good, so do it, even if the percentage of business derived from junk mail is small. Much of the time less than 5 percent of a mailing generates any business at all. When it comes to paper, the U.S. gets another badge of distinction as our 4.5 percent of the world's population uses about 30 percent of its paper. Each year the U.S. also uses about 100 billion plastic bags which require about 12 million barrels of oil (504 million gallons) to produce.

The grass clippings and leaves that so many of us have carted away as garbage are the fertilizer a lawn needs. If left to decompose back into the ground, the things that grew would replace nutrients taken out in the first place, and new growth would take place: organic fertilizer once a year is okay. A mulching mower is perfect to grind up the grass and leaves (mow half as much and save a lot of energy and pollution). Using the material as compost is also a great way to cleanly put back nutrients.

Instead, the typical approach is to cart the material away, stored in a plastic bag that caused various kinds of pollution in its manufacture and shipping, starting another chain of pollution and energy waste. This carting away of yard waste to a landfill or incinerator usually requires a truck, and even if the materials are commercially turned into compost—a good thing—the process still causes needless pollution. Then someone drives to the store to buy manufactured fertilizer and spreads it on the lawn, commonly too frequently and in quantities that are too large. Along with the poisons—many of which are carcinogenic—that we put on our lawns to kill weeds and insects, the fertilizers often percolate down into our drinking water aquifers and pollute them. A lawn does not have to be perfect in shape, texture, and color. Mine isn't and I am still alive and well, and no wildlife or pets die because of my lawn care—but nationwide substantial numbers of animals die every year from poisons on lawns. No children are affected either, immediately or in later life (acute poisoning or chronic poisoning) from ingesting the poisons directly, or from the air they breathe or water they drink. Children are much more susceptible to chemicals than adults. Household pesticide use, including on lawns where children play, may cause leukemia and brain cancer, as well as damaging motor skills and creating intelligence development problems in infants. Every year in the U.S. over 125,000 calls come into poison control agencies involving pesticide exposure and potential poisoning. Over half of these reports pertain to children.

Look at the things we use and touch in our everyday life, right down to the simplest and smallest item like that paper clip, or the action of using a paper towel or hitting a button to dry our hands, and then up to the largest items such as a car or a house. Then follow the energy chain from the item's manufacture, use and maintenance to its disposal. Then think of the number of people doing the same things on a constant basis—and the enormous number of people who haven't begun to consume but who soon will. Then be creative for a moment and think of all the different ways that we could save energy and pollution. An interesting example is the choice between using the paper towel in the public bathroom or the blower to dry our hands. One requires trees being cut down and processed, and the other requires fossil fuel burned for electricity, so try this instead. Shake off your hands, run them through your hair, wipe them on your pants or arms, and leave the bathroom. I do it all the time, and more living things stay alive because of the choice. Okay, I know you don't like this one. So maybe it's time to run that blower on solar power.

2
FOSSIL FUEL LIFE

The population growth and economic expansion in developing countries that is mounting so fast makes keeping an accurate measure of fossil fuel reserves difficult, but one thing is sure, fossil fuels are diminishing. From the first moment someone discovered we could process and burn fossil fuels, the laws of physics told us that they must run out, but the word "finite" was generally ignored. Probably the most famous prediction is indeed the most visionary, made by a petroleum geologist by the name of M. K. Hubbert. Around 1956 he went against the blindly optimistic, conventional wisdom of the time and calculated that peak oil production in the U.S. would take place between 1966 and 1971, and then the dwindling remains would be harder—and more expensive—to extract and would run out. In the 1980s we looked back and realized that U.S. oil production had indeed peaked about 1970. Defined, peak production means that we start producing less, every year from now on. When plotted, peak production is a curve starting with a steep line going up—representing more and more oil being produced year after year—then the line rolls over and heads down as every year after the peak, less and less oil is produced. The U.S. government is well aware that for all practical purposes it will very soon run out of oil. Some estimates are for about 20 years of oil remaining in the U.S., if we stopped importing oil. Others say a bit longer, and indeed it may last longer than 20 years; import availability, conservation, economic downturns, new discoveries, secondary and tertiary recovery and other energy sources will extend our limit, but regardless, the end is very near—then we will be 100 percent dependant on imports, while much of the world is in the same situation. Hubbert's name is now synonymous with the famous Hubbert curve, the graph that shows U.S. peak oil production. He also did a calculation on world peak oil production and determined that at the low end it would peak by about 1990, and at the high end about year 2000. Not surprisingly, some of today's estimates show peak world production of crude oil happening around 2005 and from there production drops off over the following 50 years. Estimates vary on this issue, but the peak is either happening now or is very near, while the world demand for crude oil keeps climbing.

As one would expect, the oil companies have been more optimistic, even with Royal Dutch Shell (one of the biggest) admitting in 2004 to overestimating their proven oil reserves by over 20 percent. What surprises do other oil companies have coming? Countries too, have been known to exaggerate their oil reserves for economic reasons, as did six OPEC nations in the late 1980s. China, Japan, the U.S.—in fact all countries—are competing for oil from everyplace in the world where it exists, as the black fluid gets scarcer and more expensive.

Some estimates of remaining world oil range from 40 to 100 years. The U.S. Department of Energy in 2004, predicted world oil shortfalls somewhere between 2021 and 2112—quite a wide spread. Considering that so much of what we need is based on oil, the dramatic price increases per barrel—despite the ups and downs in price—that have already started will continue and collapse the economy if we keep depending so heavily on the stuff and don't at least dramatically reduce our use of it through conservation. Even the governments in the Middle East know they are running out of oil. There is a saying there that goes something like this: "My father rode a camel; I drive a car; my son will ride a camel." They have little to sell the world except for the oil that is the basis of all world economies. Whether oil runs out in 75 years or 100 years doesn't matter because we are in the final chapter of the fossil fuel era, and continuing to rely on any fossil fuel is dumb, from an economic standpoint, a human health standpoint, and an environmental one. Fossil fuel use is killing us. (We may never completely run out of oil if at some point we stop using it so extensively. However, the real danger is not running out, rather, long before this, oil shortages that lead to economic collapse and war.)

In the U.S. the lower 48 states are producing less than half the oil they did in 1970. Although there is barely any oil left in the United States, and although we have only about 4.5 percent of the world's population, Americans consume 25 percent of the world's oil. The energy equivalent of only one barrel per year or less might be used by a person in a developing country (this is changing), whereas in industrial nations one person will use the equivalent of 20 to 60 barrels per year. In 2002 the worldwide average was about 11.4 barrels per person per year. In 2003 each average American consumed at the high end: the energy equivalent of about 60 barrels of oil burned per year. That's about 2,520 gallons. The numbers since then are similar.

U.S. fossil fuels are all running out, although at different rates. For instance, there is a lot of coal in the U.S., possibly a 250 year supply at current consumption levels. Furthermore, coal can be used in ways other than burning to produce electricity, including being liquefied to run a car, or even turned into a gas, collectively called synthetic fuels or synfuels, but the processes are very expensive, energy consumptive, polluting and destructive to plants and animals. Plus, consumption levels are rapidly climbing, so a 250 year supply only looks good on paper. And if we do decide to use coal to run everything, then its supply figures won't even look good on paper anymore. Deep under the crust there is more coal but it is not economically obtainable.

Natural gas (primarily methane, CH_4) is a great fuel to burn for energy, and it is much easier to get out of the ground than coal, but estimates in the U.S. are for maybe 30 to 50 years of methane supply remaining, not accounting for demand, which is rapidly increasing. Proven reserves of natural gas show less than 30 years remaining. The world's proven reserves show about 70 years remaining at current rates of consumption. Estimated reserves take the numbers higher. For North America some estimates claim over 100 years remaining. There is a lot more gas available in the form of methane hydrates, which is natural gas trapped in frozen water. These deposits have been identified deep under the permafrost in the Arctic tundra as well as in deep ocean sediments, but it is feared that mining this could cause an accidental massive release of methane, which would add to global warming. Methane is a much more powerful greenhouse gas than carbon dioxide. However, we may not have to wait for such a release because as global warming proceeds on course, many of these methane hydrate deposits may be released on their own, as the permafrost melts and the oceans heat up. In any event, although some countries are attempting to mine the hydrates, they are very difficult to obtain and put to use. Coal-bed methane (gas associated with coal) may also be abundant, but there is no rational point in attempting to mine and burn all of our fossil fuels, unless we want all the associated pollution and negative health effects—and of course, guaranteed climate catastrophes.

Let me stop here to mention a technology that has been talked about quite a bit lately. That is carbon dioxide (CO_2) sequestration, a process in which CO_2 is taken out of a pollution stream and stored somewhere, such as underground or in the oceans. Basically, when coal is burned, the CO_2 produced is trapped and prevented from being released into the atmosphere, minimizing global warming. But this process is very energy consumptive in itself, it doesn't take out other pollutants unless the process is modified, and it is very expensive to build the systems and machinery for each plant, a cost that is passed on to the consumer. One plan will inject CO_2 into the ocean depths, while another is to inject the CO_2 deep underground into the supposedly perfect geological formations that would trap the CO_2. The possibility of gradual or sudden leaks is being considered, and if leaks back to the atmosphere became common, the whole process becomes a waste of time, energy and money. Also, large scale leaks back to the atmosphere can kill people in the area by suffocation. In 1986, through a natural and unusual process, Lake Nyos in Africa overturned and released CO_2 that killed over 1,700 people as well as thousands of cattle and other animals.

We are nowhere near fully implementing this technology in the U.S.—although it is starting, it may take 15 to 20 years for substantial progress—and there is so far, no guarantee that over time much of the CO_2 won't find its way back to the atmosphere. There are potential ecological effects of releasing CO2 into ecosystems, such as changing the chemistry of sea water to the detriment of sea life.

Another problem with CO_2 sequestration is the gigantic quantities of CO_2 that would have to be stored. In Norway, one company is injecting CO_2 into sandstone beds 3,000 feet below the ocean floor at a rate of about one million tons per year, which is a small percentage of the yearly Norwegian CO_2 emissions. At just one, 1,000-megawatt coal fired plant, the equivalent of about 50 million barrels of high pressure CO_2 would have to be sequestered. In 60 years of operation this adds up to three billion barrels, which equates to 126 billion gallons! If sequestration is fully implemented, hopefully the technology will advance and allow for safe storage of the gas. However, CO_2 sequestration will delay the move to a clean, safe energy source such as solar, and fossil fuels will still remain finite. Pollution from fossil fuel use will still continue to some degree, and the habitat destruction and pollution from mining operations will also still continue. Further, the CO_2 we talk about trapping is only what comes out of power plants burning mostly coal. What about everything else that puts out CO_2, such as vehicles, aircraft, homes, manufacturing, recreation, etc.?

Carbon dioxide sequestration is being slowly implemented by the European Union and will likely spread to the rest of the world as the technology improves, but it is not yet an up-and-running full-scale industry. The sure benefit of CO_2 sequestration is monetary, for the fossil fuel companies and the politicians they support. It would be much wiser to quickly put the enormous energy, time and money that will go into CO_2 sequestration into conservation and alternative energy, so as to stop burning fossil fuels altogether.

We continue to build coal burning electricity plants in the U.S.—over 250 are now being planned or built—and China plans to build hundreds of plants, in 2007 at a rate of one to two new plants begun per week, with India also rapidly expanding and building plants. Over 850 coal burning plants may be built between the three nations over the near term, and their CO_2 emissions will increase dramatically and dangerously unless of course, sequestration that works is used. The U.S. alone emits about two billion metric tons of CO_2 a year from coal burning alone. Worldwide, from all human activities, billions of tons of carbon dioxide are sent into the atmosphere every year.

Some people advocate raising the gasoline tax to discourage overconsumption and reduce CO_2 emissions. Taxes on gasoline in the U.S. average about 46 cents a gallon (federal and state combined), whereas in Europe and Japan taxes are measured in dollar equivalents. In November of 2004 a gallon of premium gas in the United States was over $2.20 while in parts of Europe it ranged from $5.34 to $6.20 per gallon. World gas prices obviously went higher in 2007-2008; in the U.S. prices rose to over $4.00 per gallon. In Europe prices averaged over $8.00 per gallon. Then with a worldwide economic decline lack of demand brought the prices down. Raising gasoline taxes would certainly reduce consumption but would do nothing to educate the public about the dangers of that consumption. Conservation for reasons other than price has not been seriously put forth as an environmental issue in the U.S., although it is starting. For example, driving at 75 mph increases fuel consumption dramatically versus driving at 55 mph. Reducing the speed limit nationwide would save immense quantities of gasoline, as it did when we had the gasoline emergency during the 1970s and a lower speed limit was mandated. In order to conserve today, must we wait for an emergency that we know is coming?

The gigantic amount of money made in the energy business is part of the reason we are encouraged to use a lot of polluting energy. Worldwide, the electrical power industries alone produce revenues of approximately one trillion dollars. Coal companies for years have resisted any effort to burn less coal and have even come up with outlandish statements suggesting that burning coal is good for the planet. The oil companies, in the same camp, do not want consumption rates less than the approximately 150 billion gallons of gasoline burned in 2005 in the U.S. I guess then that global warming, acid rain and pollution in general are good for the planet. Would the fossil fuel industries have us believe that coal burning—which every year sends into the atmosphere tons of methyl mercury (methyl mercury is detrimental to fetuses and infants, causing developmental and neurological damage), some of which finds its way into fish and our children—is also good for us? Cleared forests and blasted away mountaintops, with overburden dumped into valleys, and strip-mining, tearing up the ecosystem, are good too? Land subsidence, sulfur compounds polluting and acidifying lakes, rivers and thousands of streams, and soil damage from replacing forests with nonnative species are good too? Tens of millions of tons of ash needing land disposal every year are also good? Then there are the hundreds of out of control surface and underground coal fires around the world. Millions of tons of coal are ablaze. These fires are accidentally caused by humans or caused by spontaneous combustion and contribute dramatically to air pollution and global warming. These fires are a global catastrophe.

Studies have shown that air pollution from burning fossil fuels prematurely kills between 30,000 and 60,000 Americans yearly and over 400,000 people yearly in China with the number expected to rise. China's leading cause of death is cancer from pollution. Several hundred thousand more people die yearly in Europe, and overall worldwide hundreds of thousands more die or are sickened. Asthma, bronchitis, emphysema, lung and heart disease, including lung cancer and heart attacks, are aggravated or caused by air pollution. Lung growth problems, as well as thousands of asthma attacks yearly in children, are induced by air pollution just in the U.S. Hospitalizations, mostly in the very young and old, also hit the thousands per year mark. In Europe the estimates are over 100,000. The associated environmental problems from mining and burning fossil fuels are many, severe, and well documented—and not only related to global warming.

In the U.S. over 55 percent of vehicles driven are SUVs, light trucks and minivans, and additionally many are vehicles with low fuel economy, such as the sexy and powerful eight and six cylinder cars so many people drive. The car companies still advertise all of these, and people still buy them. Heavy trucks too, are notoriously inefficient and are easily redesigned to get much better fuel efficiency—the redesign is underway.

Americans continue to spend hundreds of billions of dollars every year on foreign oil. The true cost of a gallon of gasoline or a barrel of oil—measured as 42 gallons—is not reflected at the pump. Before the Iraq war the cost of our military presence in the Middle East to insure the oil flows was more than 50 billion dollars a year. A rough estimate now would add about 60 dollars to a barrel, so that during 2008, when oil hit 147 dollars a barrel, the true cost would have been about 207 dollars a barrel for Middle East oil. If all the associated costs of the Iraq war are factored in the cost per barrel would be substantially higher.

We don't recognize the true cost of oil unless we understand the income tax we pay, part of which goes to our military presence in the Middle East. When the environmental damage associated with using oil is factored in, along with the damage to human health from pollution, the true cost of a barrel of oil is higher yet. When the similar associated damage from burning all other fossil fuels is factored in, we should be upset. The true cost is hidden not only in our income tax bill, but also in our health insurance premiums and medical expenses, our food costs, and manufacturing and transportation costs. If the International Center for Technology Assessment is correct in their calculations, then the true cost of a gallon of gasoline in the U.S. in 2007, was easily over 10 dollars a gallon. Burning fossil fuel is expensive and it kills people, other animals, plant life and food crops—and shortens the life of our planet.

In addition, oil spills are a common occurrence related to drilling for crude oil, not just the big ones we hear about, but the thousands of little ones we never hear about unless we look towards places like Nigeria. There, wetlands and habitats have been ravaged by the oil companies because environmental regulations are not enforced.

Just one big spill, the Exxon Valdez incident, has been estimated to have cost about 15 billion dollars and killed vast amounts of marine life. The refining of fossil fuel products is full of leaks and spills, as well as air and water pollution from refineries and distribution centers all over the world. No one knows what the real, hidden cost of a gallon of gasoline is, but again, it is a lot higher than we see at the pump.

There are other forms of fossil fuel, but their benefits might be more for the people who want to make money from these energy sources. Oil shale, a sedimentary rock containing the hydrocarbon kerogen, has been found in large quantities in Colorado, Utah and Wyoming, but to yield about one-half of a barrel of oil would require mining about a ton of oil shale, and the environmental damage from this process is considerable. And of course oil shale too, will run out someday. Also, the amount of water used in the production of oil from oil shale is tremendous, and rapidly diminishing water supplies are a major problem in the U.S. as they are around the world.

Oil sands are another source of energy being mined; Canada has the world's largest deposits. Oil sands are also sedimentary, like oil shales, and contain bitumen, another hydrocarbon. The U.S. has some of these deposits, but again, as with any fossil fuel the environmental damage from going after this is problematic. Burning these products also produces similar pollution as burning crude oil products, coal and natural gas. Natural gas is generally the cleanest burning of the three but do not miss the fact that all fossil fuel burning puts out carbon dioxide—that greenhouse gas—in addition to other pollutants. Coal tends to put out the most carbon dioxide and natural gas the least, with oil in between.

Additionally, of the total amounts of oil sands and oil shale out there only about half of it may be recoverable, and the mining and processing is extremely energy consumptive in itself, and expensive from a dollar-cost point of view.

Since the 1970s the total number of miles driven has about doubled in the U.S., while the population has not doubled. To make matters worse, we still get less than half the gas mileage easily available to us. In 1988 the average miles per gallon for the U.S. passenger fleet as a measure of fuel efficiency were estimated to be about 26. In 2004, mostly because of SUVs and light trucks, that average was down to about 20 mpg (similar in 2008). In 2004 cars averaged about 24 mpg, SUVs about 18, and pickups about 17 mpg. The true numbers are probably lower because EPA estimates of fuel economy have been unrealistic given the way that mileage tests are conducted. Over 60 years ago the Volkswagen Beetle got thirty-something miles per gallon with old technology. In the 1970s fuel economy of near 40 mpg was happening. Today, Volkswagen, Audi, Mercedes and Honda have built cars with mileage ranging from 60 to 200 mpg. Some of the cars are perfectly practical; some are not. Some are on the road such as the two-seater Honda Insight hybrid, which claims about 70 mpg, or the Toyota Prius Hybrid, reaching 55-60 mpg. A conventional 2004 Honda Civic with a 5 speed transmission has achieved 43 mpg on the highway. In 2007, the Audi A4 diesel claimed 30 mpg local and 52 mpg on the highway. The Volkswagen Lupo, a turbo-charged diesel sold in Europe, was long-distance test driven in 2000 and averaged 99 mpg.

Does the public really understand what high mileage is and what it means to us? Do some Americans simply refuse to buy a high-mileage car? Do the manufactures of vehicles refuse to fully market them? Gas-electric hybrid technology was in use 100 years ago. Modern gas-electric hybrid technology is over 20 years old. The comfortable hybrid that gets 55 mpg is already on the market. So, where have we been while so much mileage sat on the shelf? Waiting for high gas prices to make high mileage more desirable, because nothing else mattered? Well, high prices are finally starting to work. A group based in California, CalCars, has already tinkered with a Toyota Prius making it a plug-in and achieving 100 mpg. Now there is talk that Toyota may put a plug-in Prius on the market, possibly by 2009. And there is talk of pushing well past 100 mpg in the near future. (For plug-ins to really be beneficial, solar must be involved. More later.)

When it comes to oil supplies, I would listen to the geologists who don't have any political or corporate affiliation—therefore, listening to the federal government should be ruled out. The bottom line is that for practical purposes we are getting close to running out of oil, and other fossil fuels are nowhere near limitless. To let the problems associated with diminishing fossil fuels, especially oil—as well as the problems of political unrest and climate change related to fossil fuels—creep up on us without a logical and immediate energy plan in place is ridiculous. If we are as close as some think to shortage problems, then there is very little time left to transition to alternative energy sources. If the price of oil goes high enough and stays high—and if we haven't seriously reduced our use of it—the U.S. economy will rapidly deteriorate, possibly into a depression. I don't think our government or corporate machine is stupid enough to consciously wreck the economy, but they may be greedy and corrupt enough to gamble for more profits by staying on oil as long as possible. If they lose the bet, then so will we.

The 2008 recession and consequent drop in oil prices will cause further dependence on oil by removing the urgency to get off it. Regardless, there is no point in being optimistic about the time frame of running out of oil or any fossil fuel, because there is nothing to gain in waiting to change; instead we push ourselves closer to serious economic and environmental problems. And for what? So that we can burn more fossil-fuel products, making oil, coal and other industries richer? So that we can drive an SUV, build a bigger house, buy more toys, let the water run or use unnecessary electricity? So that we can use a blower to clean our walk or lawn instead of a broom or rake? The only safe way to go is to be pessimistic and assume that fossil fuels, especially oil, will run out—or become too expensive—sooner rather than later. This way the risks are eliminated, and the planet is a cleaner and healthier place.

The U.S. government has never had a serious energy policy other than to keep burning as much fossil fuel as possible. It does not raise fuel economy standards significantly, or with any immediacy, and does not educate the public to seriously and realistically conserve energy. Nor does it lower speed limits to 55 miles per hour nationwide, which would also save lives. How long will it take for serious, meaningful change? How many more presidential administrations and years before we realistically get it together?

In the United States all the easy-to-get-at oil has been found and used up, and we do not have enough to keep our economy running without continuous imports of oil. Do we really want the Middle East, or any of the other countries we import oil from, controlling our economy? Today, we import almost 70 percent of our oil. This number will jump dramatically during the next 10 to 15 years unless we do something in a hurry. For the U.S. government and the American people to have allowed themselves to get to this position both economically and environmentally is not sensible to say the least. We have been warned endlessly over the years by scientists and others about the problems associated with fossil fuels, and for the most part we ignored the warnings. Over 250 million years of fossil fuel formation is being used up in about a 150-year span.

Recall the oil problems of 1973, when Arab leaders placed an oil embargo on the United States. The Organization of Petroleum Exporting Countries (OPEC) formed a cartel that purposely cut back production to the U.S. and quadrupled the price of oil. Remember the hour-long gas lines, shortages, and the dramatic price increases for a gallon of gas that resulted? At times a 5-gallon limit per customer was imposed, and this was when we had a national 55 mph speed limit. Even decorative lighting during the holiday season was reduced to save energy.

Recall the Persian Gulf War of 1991, in which tens of thousands died and vast environmental damage was done from fighting the war as well as from the approximately 700 burning oil wells. If the world lets oil become even more important than it is now, then we face more wars. Confrontation with China is possible, especially considering that China talks oil deals with Iran, while at the same time the U.S. talks about military strikes against Iran. China is also talking oil deals with some of our suppliers, such as Saudi Arabia and Venezuela, which can lead to dangerous competition. (Recent intelligence reports show China has retargeted some of its missiles at the U.S.)

Issues will come and go, but new events down the road will continue to happen, and as oil gets scarcer and scarcer, the likelihood of a clash between superpowers will increase. War between other countries could also be the case, and let's not forget that nuclear weapons are for real. We have taken the first steps towards much greater conflicts.

In 2003 the U.S. government even went as far as to offer substantially higher tax deductions for the purchase of a Hummer (about 10 mpg) than the purchase of a 55 mpg hybrid. The Bush administration also instituted new regulations for 2008 – 2011, asking for a one—that's right, one mile per gallon—increase in average fuel economy for light trucks by 2010. The plan also exempted bigger SUVs such as the Hummer H2 and Ford Excursion from following any fuel economy standards. These new regulations were in place for one year when common sense prevailed, and in November 2007 with a push from 13 states and cities as well as four environmental groups, a federal appeals court rejected the new standards as not tough enough. How is it that a government is willing to go to war over a resource like oil and then does almost nothing to encourage conserving oil products, and even encourages wasting them?

The United States government is fully aware of the time frame for being almost completely dependant on foreign suppliers of oil. Therefore they must ensure the oil flow at all costs, even if this means military conflict. Yet keep in mind that we import less than 25 percent of our oil from the Middle East. The rest comes from places such as Canada, Venezuela, Mexico, Nigeria, Angola, Colombia and others. So why are we so hung up on the Middle East when doubling fuel economy—and we could easily triple it—along with conservation and technology could make Middle East oil completely unnecessary?

We can continue to drill everywhere possible—including the Arctic and the Alaskan wilderness, where some of the most pristine and untouched wildlife areas exist—but because of the way in which we currently operate, this will buy us little time. Alaska's North Slope, millions of acres making up the northern section of the state, is under attack. Drilling leases there run from a few hundred acres to 15,000 acres, and the National Petroleum Reserve-Alaska is a 23 million acre area being leased for drilling. Overall the North Slope may contain 48 billion barrels of oil, which at current rates of consumption would only supply the U.S. public's habit for an additional seven or eight years. Even the Arctic National Wildlife Refuge is under consideration for drilling. We are polluting and damaging too much of our pristine wilderness, with thousands of oil rigs and with the associated processing, pumping and transportation facilities needed to do the job. Habitat gets fragmented by roads; more trees get cut down and more animals get pushed onto smaller parcels of land adding to species extinction.

Then there are the natural gas wells being drilled all over Montana, Wyoming, Utah, Colorado and New Mexico. In 1998 there were about 25,000 gas wells, and by 2004 there were almost 43,000. Another approximately 5,800 new drilling permits were given out in 2004. Vast quantities of water have been pumped out of the ground to get to some of the gas, and much of this water has been wasted, at a time when water shortages abound in the west. Aquifers—nature's underground water supply—are ruined, and land is affected as habitat and wildlife are impaired.

It is estimated that the increase in U.S. oil needs over the next 25 to 40 years could go up by 65 percent. The estimated natural gas needs over the next 20 years will go up by 40 percent or more. More commercial growth, more people, more big houses and more vehicles demands ever more natural gas and oil products, with no end in sight, all for a way of life that is one of the most excessive and wasteful on the planet. How much endless growth of our economy and our population will the estimates of remaining fossil fuels allow? Will we find some more areas of the planet to ravage for fossil fuels? No doubt we will find more oil in the Gulf of Mexico and off the coast of California using deep drilling techniques, but the finds will be just an extension of time that inevitably must run out. Moreover, what of the associated pollution and global warming from all the energy used to get and process the new fuel, before we even burn it? If the rest of the world used oil like the United States does, then today the world would be using about 20 times the amount of oil that the U.S. is currently using.

The energy the world uses every day and every year is enormous. The exact amount of world energy use is a challenging number to establish, and there are varying estimates. However, world energy consumption in 2002 was about 410 quadrillion Btu's and has since increased. A quadrillion is one thousand times a trillion! (Or 10 to the fifteenth for you math people.) If we convert this amount of energy to a barrel of oil equivalent, we would use over 70 billion barrels in one year worldwide. (That's about three trillion gallons of oil.) Considering one overly optimistic estimate that says there are about 1.8 trillion barrels of oil left in the earth—not all of which is economically recoverable—then in 2007 if we obtained all of our energy from oil, world oil would be gone in about 26 years. I will repeat something. Over the next 50 to 100 years, if we keep growing economies and population as we are, our energy demands may increase by 400 percent, maybe more.

3
NUCLEAR POWER

Splitting the atom releases incredible amounts of energy. Controlled fissioning of one pound of uranium—about the size of a small book—can release about as much energy as 6,000 barrels of oil, which is the equivalent of 252,000 gallons of oil, or the equivalent of about 1,200 tons of high quality coal. In explosive terms—depending on bomb design—one kilogram of material such as uranium-235 (roughly 2.2 pounds of the stuff), can be equivalent to the explosive force of about 17,000 tons of TNT. TNT is trinitrotoluene, a common chemical high explosive used in blasting and artillery shells. The atomic bomb dropped on Hiroshima during WW II killed approximately 150,000 people (more over time from long-term radiation effects) and was only about a 15 kiloton bomb, the equivalent of 15,000 tons of TNT. It was made up of more than 2.2 pounds of uranium—in part because not all the material reacts. Taken a step further to the fusion bomb, or hydrogen bomb, one kilogram of radioactive material is equivalent to 50,000 to 100,000 tons of TNT. Remember a ton is equal to 2,000 pounds. Today we measure yields more in megatons—one megaton being equivalent to one million tons of TNT!

The reason materials like uranium and plutonium release so much energy is not because a pound of uranium has more energy in it than a pound of oil or coal. They have the same amount of stored energy. This is not a contradiction. It's simply that chemical reactions such as burning oil or setting off TNT or some other explosive do not turn nearly as much matter into energy as does a nuclear reaction. If you want to know more—not necessary here—about these interesting complexities, look to Albert Einstein, one of the greatest physicists, and hit the physics book and look for $E = mc^2$.

There is a lot of very useful heat energy released in peaceful nuclear reactions where fission takes place, and there are practically no pollutants released such as the ones that come from burning fossil fuels, including greenhouse gases. But there are problems with using nuclear power to generate electricity. One estimate of the remaining economically recoverable uranium in the U.S. to currently run its 104 reactors is enough to last about 150 years. If the U.S. converted all its electricity generation to nuclear, at today's rate of consumption, the country's supply of economically recoverable uranium would drop to about 30 years. But the definition for economically recoverable does change, as the cost of traditional resources increase, and technology improves. Therefore the amount of available uranium could increase. World estimates of economically recoverable uranium vary, but it does have limits. With reprocessing of spent fuel, which England and France engage in,--the U.S. is now considering the process—nuclear material used from dismantling nuclear weapons, and existing stockpiles of fuel, near-term supplies of nuclear reactor fuel are adequate. Of course, how fast and to what level the world increases its use of nuclear power also enters into the equation. And, England and France have considered ending reprocessing as it is very expensive. It is also a dangerous process and increases the likelihood of terrorists obtaining weapons-grade material.

The problem of long-term supply might be solved by breeder reactors that produce plutonium and can use plutonium (Pu-239) and highly enriched uranium as a fuel, some of which comes from reprocessed nuclear waste. These reactors could extend nuclear fuel supplies for hundreds or even thousands of years.

But breeder reactors are extremely complex, expensive to build and operate, and some say are much more dangerous to operate. (Note: Nuclear reactors cannot explode like an atomic bomb.) What's more, reprocessing spent fuel runs a greater risk of a serious radiation release, and there would still be radioactive waste to dispose of. A meltdown—when the reactor goes out of control—at a breeder reactor would be far more serious than a meltdown at a conventional reactor—which is no joke to begin with. There is a reason why only three breeders operate worldwide (2007), and they do not operate continuously. Germany and England shut down their breeder programs; so did Japan after numerous accidents. France has one breeder reactor, which operated for less than one year over a 10-year period. In the U.S. a breeder project at

Clinch River, Tennessee, has been around for over 15 years, is not finished, and is costing over five times the original almost two billion dollar projection. There have been experimental breeder programs and as the technology has advanced some countries are now planning breeders, such as India. (Germany and Sweden, considering phasing out all nuclear power, are planning for more renewables.)

A breeder meltdown can be explosive and rip through containment structures, and would release plutonium (Pu-239), which has a half-life of 24,000 years. This is the time it takes one half of any radioactive material to decay to a harmless element. It takes 8 to 10 half-lives to reach negligible radioactive levels. Furthermore, plutonium is the stuff that makes the best atomic bombs, and in one year, France alone can produce around 16 tons of plutonium, enough for many atomic bombs. Terrorists would love a world with plutonium everywhere to steal. The French produce most of their electricity from nuclear power and are attempting to achieve all of it from nuclear with the help of government subsidizes.

The U.S. has about 17 times the surface area of France, has about five times the population, and U.S. per capita energy use is almost double that of the French. So for the U.S. to accomplish what France has done with nuclear power would be a formidable task.

We also still produce large numbers of nuclear weapons, as well as electricity from nuclear reactor fission. These processes produce vast quantities of very dangerous nuclear waste that can be radioactive for thousands of years. There are tons of deadly radioactive waste stored in the Hanford Nuclear Reservation in Washington State and elsewhere, and there have been spills and leaks at some of these facilities. Something like 10,000 tons of radioactive waste accumulates every year from the worldwide use of nuclear power. At the end of 2006, over 270,000 tons of waste were being stored worldwide, 53,000 tons of it in the U.S., most of it on site at the nuclear power plants themselves, and some of these storage capabilities are being rapidly exhausted. Nuclear waste is piling up all over the world. As mentioned in the introduction, for years it was dumped in large quantities in the oceans by the U.S., Britain, Japan and France until the practice was banned, but the Russians were rumored to still be at it. In 1993 they dumped 900 tons of radioactive waste into the Sea of Japan. Just north of the White Sea is a Soviet military dumping ground where 17 spent nuclear reactors and submarines have been quietly dropped below the waves. These are highly radioactive devices, and some still contained live fuel. Pristine Arctic waters are already polluted to the point where PCBs (polychlorinated biphenyls) and heavy metals show up in polar bears, seals and whales and also in the human systems of the Inuit that eat contaminated animals. Similar problems have occurred around the Aleutian Islands off the Alaskan Peninsula as a result of U.S. nuclear activities. On land, the Russians have been no better, as they have pumped millions of gallons of liquid radioactive waste directly underground, straight in, with no protective containers—not that protective containers will outlast the materials in them!

In 1999 in southeastern New Mexico, over 2,100 feet below the desert floor in salt caves, the storage of radioactive waste from nuclear weapons plants and laboratories began. As of 2003, 14,000 drums of plutonium waste have been received there.

Another rarely mentioned aspect of nuclear power is that parts of the power plant itself, as well as tools and materials used in the industry—tens of thousands of tons of this stuff exists in the U.S.—eventually become large amounts of dangerous high- and low-level radioactive waste that must be disposed of. Further, the mining of uranium produces vast quantities of radioactive mine waste. In the U.S. alone about 200 million tons of this type of waste is piled near processing plants and mines. It is a threat to air, groundwater, streams and rivers. The mining of uranium and its enrichment to fuel grade can be extremely polluting to the environment and hazardous to workers.

Nuclear reactors have a given lifetime. It was thought that they would last over 40 years before having to be decommissioned and dismantled. However, around the world about 100 nuclear plants have been shut down in an average time of about 20 years. This is due in part to the complex mechanism of generating nuclear power which causes materials to be much more subject to corrosion and embrittlement. These complexities and shut-down times are another reason why nuclear power is much more expensive than originally thought. Newer reactor designs have solved some of these problems, at least helping to keep costs down.

In addition, the enormous cost of waste disposal and nuclear power in general is not completely figured into our utility bill; it's hidden. Annually, about 10 billion U.S. tax dollars subsidize the nuclear industry, and more money will be spent in the future.

One small plant alone—Yankee Rowe in western Massachusetts—cost over 360 million dollars to dismantle, and over 50 million of that money went to build temporary storage facilities for the low- and high-level waste generated throughout the process. The Maine Yankee plant, in Maine, is estimated to eventually cost 635 million dollars to decommission; the plant's construction cost in the late 1960's was about 231 million dollars. No spent fuel has been taken from the site, and it is not certain when or if the Department of Energy will fulfill its obligation to do so, although they claim an approximate 2012 date. Lawsuits by utility companies have been filed against the federal government and are estimated in the tens of billions of dollars for potential liability over breaching contracts to remove spent fuel since 1998.

There is great financial risk in nuclear power in that an operating plant can go from an asset to a liability in a matter of seconds. Basically, a meltdown is a situation in which the fission reaction goes out of control, extreme amounts of heat and radiation are generated and the reactor is completely destroyed. The large containment structures around reactors are there to contain heat and radiation in the event of an accident, but things do not always work as planned.

When a meltdown occurs the plant is finished, costs skyrocket and cleanup takes years. Nuclear power is also far from risk free to the people who work with it. There have been numerous problems relating to radioactive releases in the industry, including an incident in Japan in 1999 at a reprocessing plant where a reaction went out of control and took 24 hours to shut down. Of the town's population, 439 people were exposed to different amounts of radiation, and three workers died from exposure. Many more people die in the coal industry than in the nuclear industry, thousands for that matter; just look at the history of mining disasters. However, these deaths are over tens of years. A serious nuclear incident could dwarf them in a matter of days. There is a reason why there are emergency planning zones around nuclear power plants. The first is a 10 mile radius, sirens included, in which people could be harmed—or killed—from direct radiation exposure. A full evacuation within this radius might be ordered. The second zone is a 50 mile radius where water supplies, food crops and livestock could be contaminated with radioactive material. Additionally, as of 2005, the U. S. Nuclear Regulatory Commission (NRC) has supplied 20 states with potassium iodide tablets for their populations within 10 miles of a nuclear power plant. When ingested, potassium iodide is supposed to stop radioactive iodine absorption by the thyroid gland, reducing the risk of thyroid cancer. This does nothing for other inhaled radioactive materials, or for external exposure to radiation. With serious radiation exposure, the tablets are useless.

In the U.S., virtually all insurance policies covering liability and property, exclude coverage for any nuclear accidents. And, the federal government, under the Price-Anderson Act, limits completely any liability of the nuclear power plant to about 10 billion dollars which is covered by a pool of insurers. This is a tiny fraction of the permanent losses that are estimated in the hundreds of billions of dollars depending on the location of the accident. When the 10 billion is paid out, the balance of payments goes to the Stafford Act where the state and federal government take over and determine how much they will pay out to the public for their losses. So, the taxpayer pays again, with another subsidy to the corporate world.

Nuclear power waste must eventually be permanently stored. In the U.S. many sites have been considered. For now, a central location in Nevada about 100 miles northwest from Las Vegas, which has been under consideration for about 20 years and will cost billions, is planned to be the main storage facility for the country's nuclear waste. This Yucca Mountain facility is a gigantic underground complex, yet is not nearly large enough to contain all the waste that we would generate if we fully converted to nuclear power. The facility can barely accommodate the waste we already have prepared for storage along with the waste generated from another few decades of current nuclear reactor operation. Furthermore, if worldwide nuclear power capacity tripled as some envision, then many more Yucca-size facilities would be needed.

Yucca is being designed to contain the very hot, very radioactive waste in tremendous quantities for 10,000 years because of the long half-life of the waste. Can the geologic, meteorologic, and social stability of the area be predicted and guaranteed for 10,000 years? Some of the best and brightest are working on the project, but it is doubtful that the distant future's stability can be determined. Incredibly, it has recently been discovered that Yucca may have water problems, which could delay its licensing.

Assuming the Yucca site goes forward as planned, eventually the waste would be trucked and moved by train from all the country's nuclear facilities to Yucca's massive, deep underground storage facility. It would take years to ship all the waste to Yucca Mountain, using hundreds of train and truck shipments. Of course, don't worry, the federal government says that there won't be any radioactive waste accidents, spills, or terrorist activities on this extensive and constant movement of extremely lethal waste. And Yucca is situated near a volcano—albeit one that hasn't erupted in about 20,000 years—and active earthquake faults. In 1992 there was a magnitude 5.6 quake about 12 miles from Yucca.

More radiation can sometimes be released in an accident from high-level waste than from a reactor meltdown. It has been rumored that in

1957 or 1958 a radioactive waste accident in the Soviet Union affected tens of thousands of people.

Potential nuclear accidents could wipe out an area, not from nuclear detonations but from radioactive release from a waste accident or a reactor core meltdown (both of which can be easily caused by terrorists). There have been many mishaps at plants and waste storage facilities, military and civil, involving leaks, spills, shutdowns, fires and other accidents that resulted in radioactive contamination of the environment. In 1991, 40 percent of French reactors experienced technical problems and were shut down. In January 2006, at the Braidwood nuclear power plant in Illinois, the Union of Concerned Scientists discovered that about six million gallons of radioactively contaminated water had leaked out over several years and contaminated land and wells. In the U.S. the military has had some of the worst problems, including numerous deliberate and accidental radioactive releases into the environment, which have contaminated wildlife, soil, groundwater and air. Cleanup around the U.S. at military facilities that produce power as well as nuclear weapons has cost the taxpayers over 50 billion dollars so far and will possibly exceed 250 billion dollars—that is if everything ever gets fully cleaned up.

The most blatant releases have been by the Russian military. Long-term intentional radioactive releases into Lake Karachay in the southern Urals, have caused at least 1,000 cases of leukemia. Standing for one hour on the shore of this lake can cause a person to die weeks later. Hundreds of thousands of people were contaminated when the lake dried up and blew radioactive dust across the countryside. Karachay is the most polluted lake—maybe place—on earth. This area of Russia continues to be highly dangerous and controversial. The former Soviet Union's military and civilian nuclear practices are rife with stories of extreme contamination and attempts to cover up the events. The manufacture of nuclear weapons is full of serious problems including the dismantling of thousands of these weapons when treaties were signed between the U.S. and the U.S.S.R. at the end of the Cold War. About 100 tons of weapons-grade plutonium must be processed and disposed of, as well as about 170 tons of enriched uranium, and some of this material will be processed into power-plant fuel.

Meltdowns are not impossible. At Chernobyl in the Soviet Union in April 1986, a steam explosion and reactor meltdown released vast amounts of debris and dust in a plume of radioactivity that came down over thousands of square miles. Approximately 135,000 people were evacuated, and eventually 170,000 people had to be permanently relocated while the destroyed reactor was encased in over 300,000 tons of concrete. Major tracts of land in the area can no longer be used for agriculture and won't be available for possibly the next 100 years. In some areas of the Ukraine people still can't drink the water or milk and can't consume local fish, fruit, meat or vegetables. The radioactive fallout from the accident spread across the entire northern hemisphere, including over the United States, and heavy fallout hit parts of Switzerland, Italy, France, Norway and Sweden. Even today an area of 1,000 square miles around the accident site has been pronounced off limits and is surrounded with a barbed-wire fence. At the time of the incident helicopter pilots who flew over the destroyed reactor to drop sand on it died weeks later from radiation poisoning. About 30 workers died months later from radiation sickness, which by the way is a torturous way to die. Rumors claim that more than 4,000 workers who helped with the cleanup have died, and thousands more have become disabled. Other results from this meltdown are the 1,500 cases of thyroid cancer since the event, and some estimates for the long-term are for 140,000 to 475,000 cancer deaths worldwide. Another estimate showed only 47,000 eventual cancer deaths across Europe and Asia as a result of the meltdown. Some of the studies may have inadvertently included disease unrelated to the accident, so accuracy does become a problem, but at the time of the accident it was believed that the Soviet government downplayed the true number of fatalities. Even today, after the breakup of the Soviet Union, the truth may not be out. Regardless, reactor mishaps are devastating.

Mental retardation in newborns, birth defects, increased infant leukemia, immune abnormalities in children and thyroid cancer have all seen increased frequency in the Ukraine and other parts of Europe where the population was exposed to Chernobyl radiation. Radiation does cause cancer, even 20 to 30 years after an exposure. This is what happened at one reactor. The former Soviet Union has about 55 nuclear reactors in operation, and they do not all comply with Western safety standards. There are about 435 operating nuclear power plants worldwide, about 35 more under construction, and 104 operating in the United States.

The Chernobyl reactor had a flawed design and was much more prone to a meltdown than U.S. designs. Nevertheless, in the United States in 1979 we came very close to nuclear disaster at the Three Mile Island reactor in Pennsylvania due to human and equipment failures. The reactor at Three Mile Island experienced a partial meltdown, and

300,000 local residents of the Harrisburg area were prepared for possible evacuation. Although it is said that little radioactivity escaped to the environment, there was so much contamination within the plant that the cleanup took over 15 years. The cost for that cleanup may be about one billion dollars, equal to or more costly than building the power plant. Who pays for this in the end?

The human aspect is something else to consider. Humans are human, so rule violations and error are a part of everyday life, including in the nuclear industry. Government agencies have been caught being lax on safety enforcement at nuclear power plants. So no matter how safe the plants become due to advanced technology, and they will become safer, a mishap will always be possible. When the area is radioactive, that's it; we can't go there, maybe for decades or even a century, depending on how much and what type of radioactive material is released. Property values collapse, tens of thousands of jobs are lost, and thousands may die.

With some new technologies we might lessen the dangers of nuclear power and "might" extend nuclear fuel supplies, but this would only generate our electrical needs. If we are serious about getting off fossil fuels, then we would all have to use electricity for heating and cooking, industry, etc., and if we weren't already onto some other type of clean transportation, then we would still have to go electric and plug in for battery charge for cars and other vehicles. All this increased demand for electricity would reduce the life of nuclear power, as well as increase the amount of nuclear waste and other associated risks.

Another nuclear accident in the U.S. may never happen; however, nuclear power is an inherently dangerous, expensive and problematic technology. Radiation is not something to take lightly and is not worth the risks involved for the long term generation of our electricity. Look up at the sun, and hold your hand out to the wind, and dip your hand into the ocean currents. These three limitless energy sources are totally clean and safe, and some of the energy, like the electricity generated from roof-solar-panels can eventually be free!

4
ALTERNATIVE ENERGY

At present there is a big push by the U.S. government and the car companies to make use of ethanol, which is alcohol made from the fermentation of biomass such as corn, grasses and other organic matter. Brazil is approaching independence from foreign oil imports for transportation, and some of their cars are running on pure ethanol made from their sugar cane crops while some run on a mix of ethanol and gasoline. Even some of the trucks and agricultural machinery they use to make ethanol run partially on ethanol. However, Brazil has about half the population of the U.S., and its per capita energy use is substantially less than that of the U.S. Brazil also does not have the number of cars that the U.S. has—and has the space to grow sugar cane, a tropical grass, which can yield double the gallons per acre over corn. Other types of biomass such as organic debris or switch grass, vary in energy yield.

Ethanol production is subsidized with federal tax credits for ethanol-based fuels, otherwise the cost of ethanol would be considerably higher. The U.S. produced about four billion gallons of ethanol in 2005, which is only about 3 percent of what we burned in gasoline. The National Corn Growers Association estimated that by 2012 ethanol production will be at 7.5 billion gallons a year or higher, and Congress supported this goal. This target may be met sooner, but this would only be about 5 percent of our 2005 or 2006 gasoline consumption. This is barely a dent in our gasoline consumption, a consumption that will keep increasing as we grow our population and economy, unless we seriously conserve. Additionally, gasoline is not the only thing we burn for transportation. In 2006 the over 150 billion gallons of gasoline that the U.S. burned did not include jet fuel, ship fuel or diesel for cars and trucks where ethanol is not used. The U.S. Senate recently passed a bill requiring 36 billion gallons of ethanol to be blended with gasoline by 2022.

If corn alone were used for ethanol production, and if the U.S. devoted all its corn to ethanol production (less food!) then we would still obtain only about 15 percent of our gasoline from ethanol. Using cellulose from corn plants would increase ethanol yield. And if we used all other sources of biomass for ethanol production, according to the National Renewable Energy Lab, we could replace 50 percent of the transportation fuel used every year, but the technology and infrastructure is simply not here yet; the estimate may be overly optimistic. Using algae as an ethanol source is being worked on and could be a viable process; however, large scale production and the technology involved is not even close to ready. There are other unique methods of ethanol production being worked on that in the future might turn out to be valuable, such as cellulosic ethanol and crop waste use.

Ethanol cannot be shipped in the same pipeline that oil and gas are moved in. This is because it absorbs water, dissolves petroleum residue and so becomes unusable. The expense of retrofitting the nation's pipelines would be enormous, and one study estimated that this retrofit would not be cost effective until 30 percent of U.S. gasoline consumption was replaced by ethanol. Presently ethanol is shipped by truck, barge or railroad, sometimes costing four times what it takes to move the equivalent in gasoline. Ethanol infrastructure would need government support, which means more tax dollars.

The space does not exist in the U.S. to grow enough corn to produce enough ethanol to run just our passenger vehicles. Sure, we could cut down more trees and grow corn and other biomass everywhere possible. But any habitat loss would add to species extinction and would exacerbate global warming because trees (and really any plant) are a carbon sink, so burn them or let them rot and the carbon in them turns back into atmospheric carbon dioxide. Even if we build a house with the wood, eventually it meets the same fate, and if we don't replant the same number and similar type of trees, the carbon dioxide will not be fully cycled out of the atmosphere again. Trees also help maintain climate in other ways (more later). In Brazil the continued expansion of ethanol is contributing to deforestation because as more land is needed for ethanol crops, ranchers are being pushed from existing agricultural areas deeper into forested areas, such as the Amazon, where they then cut down or burn forests to make way for cattle. And worldwide more deforestation is taking place specifically for ethanol crops. Studies have shown this land use change to emit large quantities of greenhouse gases to the atmosphere.

It is said that the production of ethanol creates a steady state of carbon dioxide because when it is burned the carbon dioxide generated is equal to the amount of carbon dioxide that is taken back in when the biomass is regrown (photosynthesis). Yet ethanol production still produces greenhouse gases and air and water pollution because the process of growing corn (or any plant) uses diesel or gasoline in farm equipment, electricity that comes from traditional fossil fuels, and fertilizers and pesticides that are manufactured in the petrochemical industry. Also, using the fertilizers adds nitrous oxide, another greenhouse gas, to the atmosphere. In Brazil they burn some of the biomass for electricity to run the ethanol plants, and this burning too, produces some pollutants. When burned in a vehicle, ethanol is cleaner and pollutes less than gasoline but still generates some pollutants such as oxides of nitrogen.

To produce one gallon of ethanol requires a lot of energy. How much energy depends on the ethanol source and the method of production, but it can be energy negative—more energy overall going into producing a gallon of ethanol than we get out of the gallon of ethanol. With corn-based ethanol, studies have shown that not only is it energy negative, but more carbon dioxide is emitted from fossil fuels used to grow, manufacture and transport the ethanol than the ethanol saves. Also, a gallon of ethanol produces less energy than a gallon of gasoline, therefore more than a gallon of it must be burned to go the same distance that a gallon of gasoline would run the vehicle; fuel efficiency can drop substantially. Another study showed that using corn-based ethanol only reduces greenhouse gas emissions by 13 percent over that of gasoline, mile for mile driven. The debate and studies continue.

Ethanol production and use will become more efficient as the technology advances. Nevertheless, it will produce pollution and greenhouse gases, and with current methods the space problem to grow the bio-mass is real. Also, let's not forget that fresh water is a dwindling resource, and massive amounts of it are needed to grow the additional biomass and to manufacture the ethanol. This is a very serious problem as droughts in the American west, southwest and south persist. There are also large quantities of wastewater from the manufacturing process, which add to the energy cost and the environmental cost.

Then there is the food problem. As U.S. and world population continues to grow, increasing the demand on ethanol production, so does the demand for food increase. Will corn go to feed the additional people and the animals we eat, or to us to drive our SUVs? Even if ethanol is derived from nonfood biomass such as switch grass, it may then compete for the space needed to grow food. There is also a limit to agricultural yields, and again, water supplies. Ethanol production may considerably drive up the cost of food, more than it already has, as crop production represents the vast majority of the world's food supply.

The most serious reason why ethanol is not something we should hang our hopes on is the climate. The organic material that can be used to produce ethanol must grow in a stable climate. Part of our energy supply would be dependent on there being no droughts, floods or heatwaves, and on the stability of a climate that is naturally unstable and is on the way to severe instability because we light so much stuff on fire (leading to global warming). Any disruptions to the biomass supply, which have happened in the past and will happen again, would mean any remaining grains and the space to grow them, would go to the mouths of people, assuming a level of morality was maintained. A dramatic reduction in ethanol production due to climate problems would create fuel shortages and dramatically drive up the cost of fuel, and as previously mentioned, would also raise the cost of food that was already in short supply due to the climate change.

There is a place for ethanol production, and if we reduced our population and conserved, the process would be more viable. As it is, the current ethanol processes are nowhere near solving the energy problem, and because of the way we are operating—growing our population and economy—will not be a solution. But the course of action is great for the car companies and the companies that grow corn and produce ethanol. They are adding big bucks to their accounts as the government gets firmly behind the associated corporations. Much of the public (not you) buys into the latest run of brainwashing commercials telling us how green the automakers are and how green we will be if we buy a flex-vehicle that can burn a mix of gasoline and ethanol. Bio-mass is called a renewable resource, but is not always renewable because of climate instability.

Considering that about 10 days of sunlight hitting the earth gives us as much energy as all the fossil fuels stored on earth, it is incomprehensible that we haven't been seriously going after this clean, and for all practical purposes limitless, energy source. The sun's energy that falls every year on the United States alone is approximately 600 times the amount of energy that the United States uses every year. Statistics such as these have a margin of error, but it is a given that the sun supplies far more energy than we can possibly use.

Charging an electric car's batteries from the sun that strikes a home roof would be clean, free, and easy to do. Not too long ago I caught an article in the *New York Times* from a well meaning individual who came out in favor of electric cars, saying we should immediately use the plug-in type to eliminate gasoline use. In other words, plug the car into a wall outlet and charge its batteries. But given how we produce electricity currently, if we were to do this, then we would still be adding to air pollution and global warming because the electricity coming to our homes comes mostly from burning coal, natural gas and even sometimes oil. Remember our energy scenarios? Further, electricity generation from these sources as well as nuclear use vast quantities of water for steam generation and cooling, some of which is not recaptured. Plugging in to these sources would actually increase water use.

The main thing that plug-in pure-electric cars would do if used today would be to reduce local pollution, but the amount of overall pollution might go up. However, it is possible that pollution levels would remain the same, or go down a bit, depending on the power plant's efficiency and the car's design. There would be some change in the type of pollution and where it comes from: a car tailpipe, or a power company smoke stack. We could get off gasoline and import much less oil, but carbon dioxide emissions would still be produced and are significant when burning coal. If our electricity came from a nuclear plant, then we would be adding to the radioactive waste problems as well as the risks inherent in nuclear power. If we generated our electricity from photovoltaic cells (solar panels, called PVs) on our homes and industries, power-plant solar, wind and ocean energy, then plug-ins would solve the transportation problem and be a fantastic way to go. There are designs for cars that are parked under PV arrays so that while we are at work or play, batteries are charged. As already mentioned, plug-in hybrid vehicles are going to dominate the high-mileage market, but they won't solve the global warming problem enough, until they are charged from a clean energy source.

There already are electric cars that work fine, and electric motors are very efficient and simple to manufacture and maintain. In 1996, General Motors produced and leased the EV-1, a pure-electric car. It was quiet, ran and operated like any other car, went from 0 to 60 miles per hour in eight seconds and eventually got up to 140 miles to a charge. Approximately 800 were leased, mostly in California, until 2002, when GM cancelled the program. People loved the cars, many wanting to buy them or continue the leases, but GM refused, claiming the cars would not be profitable.

California set up plug-in charging stations for the electric cars—some used solar—and the oil industry helped form a group to put an end to the stations. Editorials were written saying the cars were not good for the environment. General Motors bought the company that produced an advanced battery giving greater range to the cars, but installed inferior batteries in the cars, and waited two years to upgrade to the better batteries. The battery company was sold to Texaco Oil in 2000. In 2000 the Chief of Staff at the White House was a former General Motors V.P.

In 1990 California started the "Zero Emissions Vehicle" (ZEV) mandate to clean up air pollution. The mandate ordered electric cars to be part of car manufacturers' sales. The car manufacturers and oil companies lobbied against this, and soon they started a lawsuit—supported by the federal government—against the State of California on the grounds that the state had no right to mandate ZEV. Although several other car companies had electric car programs, eventually California gave in to the lawsuit, and the electric car programs were scrapped. In 2005, General Motors crushed and destroyed the last of the EV-1 electric cars. The only remaining EV-1 is in a museum. Oil company profits in 2005 were about 64 billion dollars.

Tesla Motors in California has developed a fully electric car that hits 60 mph in four seconds, has a top speed of 125 mph, and has a range of 250 miles between charges. This means that if the car is plugged in and charging when not in use, it may never run out of electricity, and the battery-life is expected to be 100,000 miles. The two-seater is sporty and comes fully loaded. As of 2007, 700 orders have been taken for 2008, with orders spilling over to 2009. With such minimal production the car is expensive at about 100,000 dollars. However, if the company makes it and mass-production expands, the price will come down to a much more affordable level. Additionally, the company refers customers to venders that can install solar panels on a home's roof to charge the car's batteries giving the owner a truly pollution free vehicle.

With improvements in car design and packages to put solar panels on rooftops for battery charge, we could have been on our way to a logical and clean system of transportation. Of course, in 2007 GM still advertised the Hummer with what they call improved mileage at 20 mpg, and Chevy introduced the new environmentally friendly "revolution" the hybrid Tahoe SUV with 21 mpg! In 2005 the U.S. energy bill gave subsidies of over 12 billion dollars to the oil, coal, and gas industries, and less than 2 billion dollars for renewable energy. However, it would seem that pressure is mounting on big industry to do something sensible. In May of 2007, General Motors announced that they want to turn their image around and become environmentally friendly. They now claim that by about 2012, possibly sooner, a new version of their electric car will be on the road. That is if GM is still on the road. As of November 2007 they are on the verge of bankruptcy.

Plug-in hybrid technology is evolving and will very soon overtake the standard gas-electric hybrid, while bigger electric motors, better batteries, lightweight composite materials and smaller internal combustion engines will dominate as the 100 mpg—and eventually much more—car soon comes to market which hopefully, is then soon charged using solar energy. This is the road—back to where we've already been—to all-electric vehicles that are charged by the sun, wind and ocean (wind and ocean energy are technically solar as they are driven by the sun), completely clean energy sources, especially the free one soon to be on our roofs.

Solar power using PVs is old news, first developed in 1954 by Bell Laboratories. PVs that turn the sun's radiant energy directly into electrical energy today run homes, calculators, watches, Martian rovers and orbiting satellites. There are PVs for many electrical needs, including lawn mowers, pool pumps, outdoor ovens and showers, home hot-water heaters, cell phone and laptop chargers, battery chargers in general and for boat electrical needs. Developing better photovoltaic cells and batteries to convert and store the sun's energy more efficiently is already happening.

The solar industry is finally starting to explode as the world slowly wakes up to the dangers of what we are doing to the environment. Over the past 25 years the cost of PVs has dropped 20 fold and will continue to do so, as has the price of computers and other electronic devices that are far more complicated than solar technology. Since 1970 there has been an exponential increase in the use of PVs, and this will continue as they improve. There are new areas in PV development that may soon make them more efficient than traditional electricity generation. Flexible roll-out panels and even nanotechnology that stamps out inexpensive, ultra thin panels will put PVs everywhere and anywhere. PVs will be roof, wall and ground based.

With no moving parts or pollution, once installed, solar panels have a life expectancy of at least 25 years, with newer designs expected to last as much as 40 years. There are different types of setups, some with battery backup and some without, and the systems work seamlessly. A common system is one without battery backup, tied into the utility grid where excess power can be sold back to the utility. The cost of a home installation varies and is dependant on how and where the system is used, how much electricity is needed, and what federal and state tax benefits are available (at least 14 states pay for about half the cost); therefore payback time also varies. The installation might cost 10,000 to 30,000 dollars, sometimes more, and payback time can be as short as five years or as long as 10 years. However, the cost of solar installations will eventually be much less, and payback time will drop substantially, in many areas to a few years. After that, except for possible repairs, the electricity generated is free for the life of the system, easily 20 years or more. Further, when the cost of a PV system is wrapped into a mortgage it is hardly noticeable, and it becomes a tax write-off as well.

Solar energy is also used passively (vs. active use in which solar energy is focused, or stored in batteries using PVs) when the sun's energy is absorbed directly for heating water or interior living areas. Building a home or office with proper design, insulation, southern exposure and passive-solar water heaters, can dramatically reduce the energy required from traditional sources for heating and hot water needs. Sometimes the traditional heating system may hardly be used.

Solar power is used in many countries. Germany, Japan and Italy all have strong government support for solar energy, for example Germany's program to install PVs on 100,000 roofs, or their solar park in Leipzig. The German government gives such strong support to solar power that PVs are popping up everywhere, on farms, homes and industries all over the country. On homes installing PV systems is already rapidly becoming cost effective. In 20 years Germany may receive a third of its energy needs from renewables, most of it solar.

In Japan the government gave approximately 1.3 billion dollars in subsidies to install solar-power systems on 160,000 homes. The Japanese space agency is avidly working on a space-based solar-power system and plans to have a prototype in earth orbit by 2010, with additional systems in orbit by 2020. If all goes well, these high-tech devices will beam the sun's energy back to receiving stations on earth. The average intensity of the sun at high altitude above the atmosphere is eight times what it is on the ground and is available all the time. The U.S. equivalent was worked on by NASA, but the project was dropped in 2004 and replaced with space exploration—although it is being looked at again.

In Spain a recent law requires new buildings to include solar energy. The U.S. does have a Million Solar Roof Initiative scheduled to be up and running by 2010, but so far it is not working because of inadequate incentives. As of January 2006, California started a 3.2 billion dollar program for rebates to utility customers who install PVs. There have been other initiatives, and there will likely be more programs and incentives, both state and federal. Private research at universities and corporations is ongoing, and it's all a start. Yet today in the U.S. less than 1 percent of our energy comes from PVs or wind.

By contrast, Denmark gets about 25 percent of its electricity from wind powered generators, many of them offshore. Germany and Spain are starting to seriously go after wind energy and overall, Europe leads the world in wind power. In the United States wind is becoming an industry, as wind generators are up and running with more planned, but much more needs to be done.

No new infrastructure is needed with PVs. Most space on every roof of every building all over the U.S. and the world is already empty. Each building can generate a large portion, possibly all, of its own electricity for everything we need, including the charging of an electric car's batteries (the entire car body can also be designed with PVs adding to the charge). The systems could be backed up with electricity from the local power company generated by PVs, wind and ocean currents. And yes, if we forgot to plug in and charge the car, we might have a problem. Well, call a cab or a friend, ride a bike or walk, or hang around for an hour or so for a quick charge. Or, swap for pre-charged batteries. That's life in the new and clean city.

Talking about solar-powered plants, it's common to hear about the space problem with exposing large numbers of PVs to the sun. Yet it has been calculated that an area 100 miles by 100 miles would generate most of the electricity the U.S. needs, as the demand is greatest during the day. That space already exists in the Nevada desert, and for those who say we need more space, there is additional space for even more

PV arrays throughout the sunny southwest. This would mean some desert habitat loss, but when compared to the damage our existing power infrastructure does, the damage becomes acceptable. Problems of energy loss through transmission (can be much less with solar produced direct current (DC), which is converted to alternating current (AC) upon delivery) might make it necessary to spread this 10,000 square mile PV system over additional locations across the country, and if this type of project is to happen anytime soon, it would need dedicated government support. If the U.S. put the same money into this project that was put into breeder reactor research and the military to ensure oil flows, we would be in far better shape. Building a PV power plant—or a solar thermal plant, more later—is easy compared to a coal or nuclear fired plant, and can be built in substantially less time than either of the two. Solar generated electricity creates no pollution or waste. Compare this to a coal or nuclear plant that generates electricity and the gigantic amounts of waste and pollution they collectively produce.

An August 2005 article in *National Geographic* magazine cited an NYU study saying, "Panels (PVs) covering less than a quarter of the roof and pavement space in cities and suburbs could supply the U.S. with all its electricity." Some say this is too optimistic. But, in the U.S. there are about 78 million single-family detached structures and about 31 million mulitfamily structures. Seven million more units are attached units such as townhouses. There are also large numbers of commercial, industrial and retail structures. If we put PVs on all roof space and some wall space, as well as unused pavement space— such as parking lots with drive-under panels to charge car batteries as well as the grid—we could generate all of our electricity needs today, and that's without the next generation of improved PVs.

So far, from our PV power plant and PV covered roofs and pavement space, we would have about twice the electricity we need. Now throw in wind power and we have three times the electric power that the U.S. needs because three states alone—North Dakota, Kansas and Texas—supply enough wind for the job. There are seven more states that have enough constant wind to supply more energy. Then let's throw in offshore wind generators and use a greater percentage of them so that land use and habitat destruction is minimized. Now throw in ocean currents and tides spinning giant turbines and the solution is complete, because we would have more than the electricity that we currently use. At night and when it's cloudy, of course, PV output diminishes, but that is where the wind, ocean currents and tides come in along with battery-backup systems, and hydrogen storage (more later). Ocean currents and tidal movements are vast and there's a lot of kinetic energy there for the taking.

The second law of thermodynamics would prevent us from being able to use all this energy due to losses in transmission and storage. However, there is plenty of energy readily available, and the problems of the episodic nature of the sun and the wind can be—must be—overcome through better transmission and storage.

Clearly conservation—which is really changing living habits, turning off lights, lowering the thermostat, etc.—would insure the system works, especially at night, when PV output drops off. Along with increased efficiency in everything from appliances to lighting, building construction and manufacturing, energy consumption could be dramatically reduced, in which case we would much more easily have the energy needed from alternative sources. Still not satisfied? Okay, leave in place the 104 nuclear power plants, dams and bio-mass systems that the U.S. currently operates for some added security until technology advances further. Now throw in amazing technologies being worked on or yet to be developed, including PVs that work more efficiently on cloudy days, or concentrating the sun's energy to heat a fluid, make steam and spin a turbine to generate electricity, and we are on the way to carbon-free energy. This all can be done. Much has already been done in countries all over the world. In fact, in Kramer Junction, California, a massive solar thermal facility generates 354 megawatts of electricity at peak output using parabolic mirrors that focus the sun to heat a fluid as explained above. This facility has been operating since 1989. In 2007, south of Las Vegas, Nevada Solar One, another solar thermal facility with a capacity of 64 megawatts, started generating electricity. However, exceeding carrying capacity, increasing population and expanding economies, and so increasing energy needs will strain or drain any system. And, our energy problem is greater than it should be because in 2008 the Bush administration banned for about two years, all solar power projects from federal land until environmental studies can be done. This while 130 proposals from companies to build solar plants on leased federal land were on the table.

Innovative energy may not be a desired outcome for at least part of the business community. The electricity from our roofs would eventually be free. Unless the coal, oil and power companies run the new, clean, safe power industry—some of them might—then these companies would suffer. However, a whole new generation of industries would be spawned in the manufacture, installation and maintenance of PV cells, wind generators and other technologies. In fact, this new generation of business has already begun.

Solar power may already be cost effective. If we figure in the reduction in health-care costs and environmental costs, savings are already in our pocket. Less fossil fuel burned for electricity and transportation means fewer hearts attacks, asthma, cancer and premature death from polluted air. It means less acid rain and global warming, so fewer dead fish and other animals, and fewer dead or damaged trees, plants and agricultural crops which alone would save billions every year. Solar power would mean less mercury from burning coal in the environment—as in lakes and fish—and so less mercury exposure to our children. It would mean less environmental damage from mining the coal, gas and oil. These are all reasons to get solar-assisted gas-electric hybrids on the road to increase mileage even further as we transition to the all-electric car. Also, what about the peace of mind knowing that we are doing something so important and are becoming more energy independent?

With some fine-tuning of the batteries and the PV-cell output, solar-generated electricity and its storage will progress rapidly, much more rapidly with government support. There are many benefits to using the sun for our energy needs. We would not need any more Middle East oil, and this could radically change our political and military strategies towards this part of the world. Significantly diminishing the importance of oil would have far reaching effects. Solar power will become much more tangibly inexpensive in terms of out-of-pocket costs if we start the mass-production process now.

Hydrogen as a fuel is an interesting source of energy. It burns cleanly, emitting only water vapor and possibly some oxides of nitrogen. But hydrogen is extremely light, and quantities of it would leak to the upper atmosphere during its manufacture, transport and use. It has been speculated—but not proven—that this hydrogen might damage the ozone layer and add to global warming through complex chemical reactions in the atmosphere. It's also volatile, especially compressed in a tank (sometimes 5,000 to 10,000 pounds per square inch). It is also difficult to transport through pipelines in part because of its low density, so generally it is compressed which requires using more energy. And hydrogen has less energy in it when compared to equal pressures of natural gas (CH_4, methane). Liquefying hydrogen allows greater quantities to be shipped but the process uses large amounts of energy and the temperatures involved are very low. Hydrogen also burns more readily than natural gas and burns invisibly, making it more dangerous. This would create the need for sophisticated leak detection equipment. The transport and delivery of hydrogen could be dangerous and getting rear-ended might take on new meaning. New technologies for the efficient use of hydrogen are being worked on.

But possibly the biggest problem is that so far the energy used to make hydrogen substantially exceeds the amount of energy obtained from the hydrogen to drive a car or run other devices (energy negative). There are different ways to produce hydrogen, but for large-scale production options are limited. One way to obtain enough hydrogen is to split water molecules (H_2O) to strip the hydrogen off the molecule. This process requires enormous amounts of electricity, and if we are going to get that electricity from coal or nuclear plants, then the whole process defeats the purpose of using hydrogen in the first place. Hydrogen's many issues, including fuel cell durability and lifetime problems, would make the use of it very expensive, more so than using gasoline for transportation.

Hydrogen can be burned, so possibly could replace natural gas, or can also be used in a fuel cell in which the electrons are stripped off the hydrogen and used as a source of electricity which could run an electric car the same way that solar-charged batteries would. Fuel cells strip electrons off the hydrogen atom for use, and then combine the hydrogen and electrons with oxygen, usually from air, which produces water. The method works: nothing is burned, and there is no pollution to speak of. Fuel cells could be used to produce electricity for just about anything, including a home. Nevertheless, the hydrogen must still be produced, transported and distributed. There are hydrogen fuel cells producing electricity right now, but the cells are still expensive—sometimes subsidized by state and federal agencies. On the upside, a fuel-cell powered car can be very efficient, pertaining to the useful energy obtained from a given amount of hydrogen.

The amount of money, time and research that will go into ways to manufacture hydrogen efficiently, as well as to build the hydrogen storage and distribution systems, is enormous. There is talk of refining solar power to produce the electricity to make hydrogen—and it has been done on a small scale—but the obvious question is, why not go straight to the solar source for our energy needs? Splitting water molecules using electricity from the sun is not a bad idea because the energy would then be stored in the hydrogen for later use to produce electricity, especially at night, and on cloudy and overcast days. The sun's limitless energy could make hydrogen production inexpensive and practical. PVs, wind and ocean currents—again, all solar—generating electricity to make hydrogen will work, and that is all the more reason to make these systems more efficient, given the amounts of energy needed to make hydrogen. Iceland is moving towards hydrogen use because fortunately for them they are situated on top of a magma body that gives them easy geothermal energy for many uses including producing hydrogen. Other areas of the world can use geothermal energy, too.

Today most of the hydrogen manufactured comes from natural gas (CH_4, 1 carbon atom, 4 hydrogen atoms), in which the molecule is split to extract the hydrogen. It has been suggested we could get hydrogen on a large scale from natural gas or coal. This would make our hydrogen supply reliant on fossil fuels that will run out, as well as adding carbon dioxide (greenhouse gas) to the atmosphere as the carbon atom gets lost and hooks up with some oxygen. Sometimes more CO_2 is produced this way than from burning the fossil fuel. Further, enormous amounts of energy still must be used in the process to produce hydrogen from fossil fuels. Changing to the new hydrogen technology and making it fossil-fuel dependant is simply nonsensical.

Overall, electric motors are simple and far more efficient than a gas-driven automobile. There are fuel-cell powered cars and buses on the road now—for example, buses in Vancouver, Chicago and 10 European cities. Honda and General Motors both have prototype vehicles operating, but the cars are so far too expensive. Some have cost a million dollars, although costs are coming down. In 2007-2008, Chevy claims they will put 100 test vehicles on the road. These SUVs will be hydrogen-powered fuel-cell vehicles and the fleet will show up in New York City, Washington D.C. and Los Angeles. In April 2007 Ford showed off its new fuel-cell vehicle, the Hy-Series Edge. It is a hydrogen-fuel-cell-electric vehicle that uses a plug in for battery charge to run the electric motor for local driving up to about 25 miles. Then the hydrogen fuel cell generates electricity for extended driving. This vehicle is still dependant on fossil fuel because it plugs in, and as already mentioned this creates problems unless the electricity generation comes from solar. Ford says the car will be ready in 10 years, but will the infrastructure for hydrogen be in place in 10 years?

Some estimates are for 20 to 30 years before hydrogen and the vehicles that run on it can be produced economically, from a monetary perspective as well as an energy one. However, as with any new process, a change in mindset and technology certainly could shorten the time frame. But until the high cost of hydrogen production and lack of infrastructure to deliver it are overcome, fuel-cell technology will remain a small scale operation. Further, by the time a hydrogen economy is ready to go, 100 mpg plus cars, and then pure electric cars could easily be mainstream. So could solar in general for all our energy needs. In which case the need for hydrogen may be eliminated to a large degree. Of course this assumes corruption gets out of the way. The U.S. government needs to get serious about the era of alternative energy—an era that I believe they have been delaying on behalf of their corporate sponsorship.

The time needed to cycle most gasoline vehicles off the road may be 10 years or more. If another energy source is not available by the time the vehicle is sold or junked—and recycled—then people will be forced to buy gas-burning cars again. This is one reason why greenhouse gases will continue to increase. It is vital that through this transitional period all vehicles are getting the highest mileage possible. However, we keep increasing population, expanding economies and putting more vehicles on the road, so even if vehicles become more efficient, large amounts of pollution—including greenhouse gases—are still produced. We need to feel a sense of urgency in getting on to new technologies. We are not moving nearly fast enough and have delayed the process much too long.

5
CORPORATE DICTATORSHIP AND MORE

Solar power would solve our energy-related problems, but there's a catch. As long as Political Action Committee (PAC) money and lobbyists—there are thousands of them— control so much of the United States government, the corporations will keep pulling the strings. So far it would appear that some corporations may have prevented solar from being pursued with any great urgency, but thankfully this is changing, as independent corporations go after the sun.

It has been hypothesized for a long time, and recently published, that the big oil companies have suppressed knowledge about global warming for many years in order to keep us using oil and gas. An ABC News program in March 2006 exposed oil-company correspondence showing executives specifically talking about convincing the public that global warming isn't real. It has been alleged that for many years oil and coal companies have been bank-rolling any global warming skeptic they could find or create to the tune of millions of dollars. (The U.S. government has also engaged in disinformation; more later.) According to the Union of Concerned Scientists, the Alliance of Automobile Manufacturers, a lobbying group for many of the world's auto makers, has a history of blocking or slowing any progress on safety, vehicle emissions and fuel economy. The number-one goal of all these related industries is for us to keep burning a lot of fossil fuel and so keep the money flow coming while the oil and coal flow keeps going. In 2007 the Bush administration blocked California's movement to require the automakers to substantially reduce greenhouse gas emissions from new cars and light trucks.

The Kyoto Protocol—negotiated in Japan in 1997, when representatives of 166 nations gathered to set up regulations on reducing carbon-dioxide emissions and thereby limiting global warming—failed completely by a vote in the U. S. Senate of 95 to zero. Not one U.S. senator would vote in favor of trying to minimize global warming gases! According to a NOVA/Frontline documentary, this vote result was at least in part because industries sent representatives to the senators with the message, vote for Kyoto and you just committed political suicide. No more campaign money, and you are out of a job. If you or I did what some of our elected officials do, that is take PAC money for favors later on (it's sometimes called a bribe), we'd be in jail breaking big rocks into little rocks with a heavy sledge hammer.

Does the United States have a democracy or is it a corporate dictatorship with a dash of democracy thrown in? When was the last time we were asked to vote on any of the hundreds of spending bills? We have little control over our government or our tax money other than through our elected officials, but too frequently they don't do what they said they would do, or they don't tell us what they are doing. Unfortunately, our tax dollars go to an amorphous land of nonreality where we don't ask for enough accountability, and so many things are not accounted for. So, how do we control decisions on the environment, assuming they are even seriously and accurately presented? One way is to use the part of our democracy that still allows us to make a change: we are free to speak, write and protest, to join environmental organizations. We can demand change.

The technology for internal combustion engines is over 100 years old. Put gas in a cylinder and light it on fire. Why are we using this antiquated technology at a time when moon walks, space walks, lasers, computers, biotechnology, electronics, nanotechnology, unbelievable surgeries and medicine and satellites have all advanced and are still advancing? We can split the uranium atom—old technology—and produce electricity to light up an entire city, or to vaporize an entire city at temperatures approaching that of the sun (I'm not saying we should switch to nuclear). Why are we still lighting fuels on fire for transportation, heating, cooking and electricity? The answer is money and special interests. We are living in a realm of technology that is about to take us into new and unimaginable areas including, if we want it, solar power for almost everything.

During WWII we were scared to death that Hitler would get the atomic bomb first and wipe us out. We were scared to death that Japan would attack the mainland of our country. We were scared to death that we would have to invade Japan at great cost in human life. So we very quickly built an entire city called Los Alamos in the desert of New Mexico and started the Manhattan Project. The United States government put together the most brilliant physicists in the world and wrote a blank check to all involved. They got anything they wanted, but they didn't have computers like we have today, or even hand held calculators, since these were not invented yet. Slide rules, pencil and paper were the order of the day. Everything electronic still ran on old style tubes in metal cabinets, and plastics were hardly even around. Color television was a dream, and modern air-conditioning was also on hold. Landing on the moon was science fiction. The idea of splitting the atom for a bomb was still hypothetical on paper. They didn't have the technology to split the atom, yet assumed new technologies would be created on the way through the process of attempting to split the atom. There was even a theory that a chain reaction would destroy the atmosphere if the bomb worked. From Tennessee to Washington State to New Mexico the entire country was mobilized for the gigantic industrial project. After a few years, on July 16th, 1945, the desert sky lit up like the sun, and we entered the nuclear age.

Why aren't we as afraid of the death of this planet as we were of Hitler or the Japanese? Ecosystems collapsing is a far deadlier and more permanent peril. If our enemy had taken over the United States, the experience would have been horrible, but we could have always regained ground and won out in the end. If this planet's life support systems break down enough, life as we know it will end. No come back, folks!

We need to push for a government and corporate structure that will do what needs to be done. Big Brother needs to be a Good Brother and put the crisis-driven power and speed of the government behind the environment. Others have said this before me, but we need another Manhattan Project to find better sources of energy. However, the project should be a giant research-and-development facility for all aspects of the environment, including direct solar power, wind power and ocean power, and we need the project yesterday. We need a Manhattan Project for the existing technologies that already work, to improve them to far greater levels, and to create new technologies in amazing areas such as nanotechnology and space-based solar energy.

We need this Manhattan Project also as a think tank, to devise ways to subsidize private industry with ideas, and supported with billions of dollars in payouts and tax-breaks so that the cost of energy technologies comes way down. A drive to put photovoltaic cells on every available roof and wall of every building in the United States must be a top priority, as well as setting up wind farms on land and at sea. Increasing energy efficiency on appliances like refrigerators, televisions, air conditioners, lights and electric motors (on everything) should also be an issue. An ongoing program to educate the public through public-service television commercials, just like is done when they want us to know about AIDS or driving while intoxicated, could rapidly inform the public about any and all environmental issues as well as conservation techniques, of which there are so, so many. Making basic, conceptual environmental science mandatory in all schools from elementary school through college would create an environmentally aware public much more willing to accept change.

Another department could work on improving construction techniques and materials for homes and industries, as well as upgrading existing structures to levels far above what has already been accomplished. Combined with increased appliance efficiency, a home's energy use could be reduced by two-thirds that of the typical home of today. Dramatic energy reductions have already been accomplished with experimental homes.

Another department in this new Manhattan Project would get the public rapidly out of large wasteful vehicles and into solar-plug-in gas-electric hybrids and eventually to purely solar-powered electric cars. Today we are light-years ahead of the primitive 1940s. If the same urgency was put into solar power as was put into the development of the atomic bomb, then I believe it would take about 10 or 15 years, and then cars that run off the sun would be available to all of us, and so would most of the heating and electrical needs in our homes and industries come from the sun.

In 1960 when the world's technology was still incredibly primitive compared to today's, John F. Kennedy was elected president of the United States. During a speech on May 25, 1961, he set a goal of going to the moon by the end of the decade, only nine years away, which at the time critics said was ridiculous. Less than nine years later, in July of 1969, Apollo 11 landed on the moon, and Neil Armstrong stepped out and went for a walk. Apollo 11 did not have the computing power of some of today's desktop computers on which kids play games. If we want solar-powered cars in 10 years we will have them. There can be a day when we pull into an electric station instead of a gas station, and if we don't have time to wait for a charge, then we will swap out our car batteries for precharged batteries and be on our way. One company is already working on this concept. If we want solar powered everything, we will have it. If we want wind power and new, clean, innovative energy technologies worldwide, we will have them all.

Conservation and increasing efficiency are hundreds of times less costly than building new coal-fired power plants and nuclear power plants and accepting the health related expenditures and environmental damage and monetary costs related to these things. Yet there are already plans to build many more coal and nuclear power plants to generate electricity, which will cost billions of dollars up front, along with the associated environmental damage, health damage and waste disposal problems all adding to incalculable dollar amounts down the road. The U.S. government has not yet completely leveled the playing field; they are not doing enough of what is needed, and so easily accomplished, to get to clean, unlimited energy. With Barack Obama and a new administration coming in 2009 there certainly will be change, but will it be enough and will it be in time?

Many countries have put into effect the Kyoto Protocol, which strives to reduce greenhouse gases, at least trying to temper global warming. (The chapter on global warming will clarify the greenhouse effect, global warming, etc.) Unfortunately, the commitments of Kyoto only run until 2012, and there is at present no enforcement mechanism, so it is doubtful that targets to reduce carbon dioxide emissions (only one of the major greenhouse gases) will be met. At a 2005 meeting of 156 signers of the Kyoto Protocol treaty in Montreal, the United States—the world's largest emitter of greenhouse gases—still refused to ratify the protocol, and the U.S. representative walked out of the meeting. As of 2007, 175 parties—including the European Union—have ratified the protocol while the U.S. still has not. To date the United States has not put in place any meaningful form of conservation. This will change by forced necessity—and by industries that profit from solar.

Unfortunately, China and India both refused to agree to limits on greenhouse gases after the treaty expires in 2012. This is a very sad statement about some of our leaders' views about the obvious environmental problems that we face. Sadly, the Kyoto Protocol's emission targets are far below what is needed to seriously mitigate global warming. Even if all nations involved met the targets, carbon dioxide emissions would continue to rise—but it was at least a first step. So far Kyoto has done little other than make news headlines and it won't do much until there is further discussion and change.

In December 2007 the nations of the world met in Bali to discuss climate change and modifications to the Kyoto Protocol, and what the future holds after 2012. Reductions in CO_2 emissions were discussed but it was said that the U.S. again blocked progress on agreements. The Bush administration said that Europe is moving too fast in its efforts to find a replacement for the Kyoto treaty.

To put in perspective how childish and dangerous many of our leaders are being, look at what scientists are saying. The average world temperature has risen by about 1.4 degrees Fahrenheit (.8 C) in the short time since about the mid-1800s, a rise caused mostly by industrialized, modern humans burning fossil fuels, which release carbon dioxide,(also by releasing other greenhouse gases). Before the industrial revolution kicked in during the 1800s, for centuries the average carbon dioxide level in the atmosphere was approximately 280 parts per million (ppm). To play it safe, to limit the world's average temperature increase to no more than 3.6 degrees Fahrenheit (2 degrees Celsius, which might still be catastrophic) would mean keeping carbon dioxide levels at or below about 400 ppm. Climate sensitivity—that is, the average worldwide temperature increase caused by a doubling of CO_2–is thought to be somewhere between 1.5 and 4.5 degrees C, maybe higher. Currently carbon dioxide levels are at about 385 ppm, roughly a 38 percent increase from pre-industrial times, caused by human activity. Many scientists now accept the unfortunate reality that we can't stop the move to 400 ppm and above.

At the rate we are going, unless we radically change, it is possible that we will double preindustrial levels from 280 to 560 ppm in this century. It is also possible that we will more than double this. A 450 ppm level is thought to be a dangerous threshold point—which may be reached in 30 years or less—that could cause an average global temperature increase that will be catastrophic: nobody knows for sure. There are positive feedback mechanisms that can rapidly push greenhouse gas levels and global warming to such an extreme that life as we know it will end (more explained later). Does anyone really want to gamble in this kind of poker game? Apparently they do. The temperature in Alaska, Siberia, and northwest Canada has already risen by five degrees Fahrenheit in the summer and 10 degrees in the winter. Winters above 40 degrees north latitude have shortened by an average of two weeks. The Arctic has warned us, as it melts and species there suffer.

Glaciers are melting worldwide, and the ocean conveyor system is possibly being affected. Now it seems that the effects of global warming are much worse than was previously thought.

A *New York Times* article from April 30, 2001, read, "Dick Cheney—current Vice President—promotes increasing supply as energy policy. Favors nuclear plants. Seeks development of oil, gas and coal and sees no easy answer in conserving." Bush and Cheney have recently suggested that conservation is not a bad idea but have done little to promote the idea.

Transportation by cars, trucks and planes uses about 70 percent of U.S. oil and produces about one-third of our carbon dioxide emissions. Simply driving slower would reduce car and truck fuel use dramatically. Switching to current high-mileage hybrid automobiles (which will be even more improved in the near future), or simply switching to conventional high-mileage automobiles could cut passenger-car gas consumption in half. If everyone then car-pooled, with just one additional person in each car, gas consumption could be cut in half again. How difficult would this scenario really be? Even if we didn't carpool all the time—it isn't always practical—the gasoline savings and oil import reductions—and the health and environmental benefits—would be phenomenal. Add in only the conservation techniques talked about in this book, and coal, oil and natural gas use would plummet. Conservation is the easiest, fastest, most inexpensive way to reduce energy consumption.

Conservation and efficiency would put the oil, coal, mining and associated companies at an economic disadvantage. Yes, some might go out of business. However, hundreds of billions of dollars are spent on coal, oil, natural gas and fossil-fuel derivatives like gasoline. We light all these things on fire for our energy needs, just like a cave-person from 100,000 years ago did with wood for energy needs. Burning fossil fuel is a gigantic drag on the economy, an environmental and health disaster for human beings and other animals and plants. Hundreds of billions of dollars would be saved by not using fossil fuels, and all that money would be freed up to go elsewhere.

Our new Manhattan Project and think-tank could tackle all environmental problems—including preparations for the possibility of new disease and pandemics. The project could also find ways to stop population growth in the world, as well as in the United States. It could also find ways to bring our economies to zero-growth and allow them to still function. This would also stop the growth of suburban sprawl which is especially important because so many home developments are built in ways that require us to drive too far to get everything we need. This distance destroys the feeling of community and adds dramatically to energy use, pollution and habitat destruction. The new Manhattan Project would also tackle the issue of poverty in places like Africa and other developing areas by educating and bringing people out of poverty in environmentally sustainable ways.

One gigantic Manhattan Project of technological research and development—and of thought process and intellectual development—would begin to solve the problems. An immediate tax on gasoline would provide enough money for the project. The money saved on not having to engage in coming resource wars like Desert Storm would cover the cost. The cost today of going into Iraq is more than 600 billion dollars, some say much higher, and the cost is escalating. A reduction in our overall military budget, which is hundreds of billions of dollars, especially in the area of nuclear weapons of mass destruction—we really don't need to destroy the world a hundred times over—and the money saved from maintaining a military to make sure oil flows from the Middle East would do it. The billions saved by stopping immigration to reduce our population levels might do it. Moving funds from projects to put a few human beings on Mars while our own planet is falling apart would help. If I am wrong about any one of these statements, I am not wrong in saying that if we did all these things, the money available would be more than enough. There is plenty of money out there if we would stop wasting it. Even without spending reductions, when we recognize a threat, money is instantly raised by taking on more debt. In 2008 with an economic crisis 700 billion dollars was generated this way, practically overnight—with more on the way—to bail out the banking system. The key is, recognizing the threat!

We don't really need the vast quantities of oil, coal, natural gas and nuclear energy that we use. We have blindly chosen to need these things, and so has the rest of the world. Europe and Japan are much more efficient than the United States, but they still have a very long way to go. As already mentioned, the United States has no serious energy policy and never has, other than to keep burning—but this will change. No presidential administration has ever presented real conservation and efficiency as an answer. Although, Jimmy Carter (1977-1981) was way ahead of the pack but he was laughed at and made fun of in newspaper cartoons because he suggested lowering our thermostats to reduce coal use and pollution. President Carter also equipped the White House with solar panels which Ronald Reagan (1981-1989) removed upon entering office.

A *New York Times* article stated that the Bush Administration was caught censoring vital data on global warming from an EPA science report before the report got to the American public. In its original form the report supported the conclusion that global warming is a serious threat to humans. The editing was exposed on the television news program *60 Minutes* in March 2006 and showed extreme corruption and indifference from the Bush Administration, not only to science, but to democracy. Entire paragraphs and sentences in the report were crossed out. A top scientist was threatened with termination if he told the truth about global warming. As previously mentioned, ABC News did a follow-up special, backing this up and showing that for over 15 years the oil companies have been trying to convince the public that global warming is not real, a campaign that was helped along by the Bush administration. This administration has been editing scientific data in general, essentially altering data to what the administration prefers. An article in *Scientific American* from October 2005, critiquing a book called *The Republican War on Science* points this out. (I wouldn't care if the title was *The Democrat's* or *Independent's War on Science*.) Warnings as obvious as the levees needing to be rebuilt around New Orleans because of the very real possibility of losing the city in a large hurricane were also ignored, as the budget for a project to rebuild the levees was cut. Welcome, Hurricane Katrina. And not only the government, but other corporations have been manipulating data. An article in Science News from September 2008, discussed the book, *In Doubt Is Their Product*. Manufacturing scientific uncertainty, and lawyers turning to mercenary scientists to generate this uncertainty, has grown into a lucrative product defense industry says the author. Environmental issues as well as product safety issues (cigarettes, heavy metals, pesticides and other chemicals, etc.) have been manipulated for a long time.

There are many other accusations against the Bush administration pertaining to data manipulation on abortion, stem cell research, condom use, endangered species, as well as to relaxing environmental regulations for industry—like coal and oil—across the board. As revealed by the Union of Concerned Scientists, during the past five years, up to 2007, 150 scientists have reported over 400 incidents of being pressured by political appointees to distort, manipulate or suppress their work on climate science. The point is that when it comes to this planet and its environment this administration has been, extremely dangerous. If not for some in our system of press and television—which I am not always fond of—bringing issues to light, would we be able to hold onto what remains of our democracy? How would it continue to evolve? And what has our government done? Did it attempt to hide data that is fundamental to the understanding of a process that may threaten the full survival of the human race, so that industries could make more money? What might be as dangerous, or more dangerous, is that our society allows this type of behavior, almost accepts it with a shrug of the shoulders.

The Bush administration did come around, but at a slow, calculated pace. Advocating conservation, raising fuel economy standards and looking at new technology were talked about. During an April 25, 2006, speech President Bush declared—as though the idea were a revelation—that we need to look at hybrid cars, ethanol, bio-diesel, hydrogen, and conservation and repeated that we are "addicted to oil." This is old news, yelled out by scientists and environmentalists for decades. Bush gave no details of a plan, other than that money and tax breaks would be used. A telling part of his speech came when he mentioned 1.2 billion dollars should be allocated for hydrogen-power research over five years, which is not enough, and only 31 million dollars should be allocated for advancing battery technology. Batteries are a major ingredient to making solar power more functional, especially in cars. Something like 12 billion dollars a month goes to the Iraq war, and billions more to questionable foreign aid that is not monitored and ends up being wasted. Tens of billions of dollars go to immigration related issues every year. We have an overall Department of Defense budget of over 400 billion dollars, and a 2007 projected federal budget (a lot of waste) of more than three trillion. The U.S. government is not serious enough about hydrogen or solar. However, the administration leans toward supporting hydrogen to satisfy certain corporations' needs to sell us something down the road.

On January 23, 2007, during the State of the Union speech, the president briefly mentioned the environment. He said we must conserve, reduce dependency on oil, drill for more oil in the U.S., look at new technologies, etc.: an echo of past speeches, more rhetoric. The indifference to scientific data and the lack of understanding of earth science from this administration has set the United States and the world so far back that we may not be able to catch up in time to stave off economic as well as climate catastrophes, or even another resource war. In 2, 8, or 20 years what thought process, party and administration will be in place? Will lessons have been learned? Will we be in a better position? Will we have done what needs to be done in time? I am not trying to time-stamp or date this book by dwelling on the Bush administration or any political party or event. I am only using this period in time as an example of where we have evolved as a society, and where we may go.

I do not care who is in the White House as long as the show is run with common sense, and logic, and without corruption. The blame goes to all the past presidencies—especially during the last 30 years, when the environmental data was piling up collecting dust—for putting us in the position that we are in today. Our society does not have a technology problem; instead it has a corruption, ignorance, and greed problem.

In 1827 a French scientist recognized the greenhouse effect. In 1863 it was again shown that atmospheric gases keep the planet warm. In 1896 a Swedish scientist, Svante Arrhenius, predicted that burning fossil fuels would produce global warming. In 1938 another scientist, Guy Callendar, suggested something similar. In 1958 Charles Keeling started a worldwide project to monitor atmospheric carbon dioxide, a major greenhouse gas, and the project continues today. Although not all scientists believed in global warming, the concept of greenhouse gases was and is indisputable. The famous film *Soylent Green*, produced in 1973, portrayed the future earth as devastated and overpopulated, with ecosystems and human societies in complete collapse. As the first scenes open, the main characters played by Charlton Heston and Edward G. Robinson are having a conversation. They refer to the greenhouse effect as having ruined the earth. For over 30 years many scientists have been warning us about global warming and climate change. As better data came in, for the past 20 years, and especially during the last 10 years, the majority of scientists became convinced that we have a problem and have been trying to convince the world. The Intergovernmental Panel on Climate Change (IPCC), an organization made up of top scientists from all over the world, stated back in 1995 that "The balance of evidence suggests a discernable human influence on global climate." Television programs, books, magazines and professors in college classrooms explained environmental problems. Very few would listen. Now that *60 Minutes, ABC News, Nightline,* PBS, CNN, *Time* magazine and many others have all in a span of months stated that we are seriously threatened by global warming and that the science is actually true, the world and the American public have started to wake up. Now television programs routinely cover many different topics pertaining to planetary damage.

Now we are listening because the climate is changing, the earth is in trouble, and humans are threatened. A recent study shows Antarctica (the South Pole) is now showing signs of stress, as ice breaks up. There is a lot more ice in Antarctica than there is in the Arctic and if it starts to melt in any severe way, sea-level rise will be cataclysmic. In 1995, 2000, 2001 and 2002 vast ice shelf break up and record size ice bergs were recorded. Animals such as penguins and krill are starting to experience survival problems. Krill are just above phytoplankton in the marine food chain and are vital to the existence of marine life. The situation somehow had to become fashionable, or maybe very scary, for ears to perk up. Now we are facing planetary change that is irreversible. How unwise have we been?

We have not been seriously going after solar power because the special interests, such as the car, coal and oil companies, haven't figured out how to put a cash register on the sun. Think about this. Imagine we can drive a car out of the dealership and never, ever go for gas. We live in homes where most, if not all, of our electricity is free. The powerful corporations do not like this, but they may soon be forced to accept the scenario. Corporate fossil fuel reality has been looking something like this:

We continue our use of gasoline and other oil products, while slowly increasing mileage and efficiency, as well as population and consumption, and therefore use as much oil as possible to make the oil companies and associated business as much money as possible. A few oil companies claim they are working on solar—and they are, slowly, conveniently without the help the government could offer. Then with the support of the federal government other corporations go out of their way to develop ethanol, make a lot of money, and allow the oil companies to continue making a lot of money, and overall there is an appearance of being "green" and saving the world. The car companies like ethanol because little change is needed for a car engine to run on it. The oil companies are okay with ethanol because they know we will never use it exclusively, and it's mixed with gasoline, so they still get to sell the stuff. Now, again with a bit of support from the government, the car companies slowly start to develop hydrogen-powered cars. Why? Not only because hydrogen is clean to use, but also because the gas stations can be converted to hydrogen stations and the oil companies can still sell us something. The oil and car companies may not be strange bedfellows.

People get cancer and have asthma attacks (especially children) from breathing the air. Heart attacks increase, premature deaths occur, global warming threatens our existence, acid rain is still happening worldwide, and trees, crops, lakes and animals are damaged or killed all over the world, all from using fossil fuel. Are these not threats worth attacking with another Manhattan Project? If we fully develop PV, wind and ocean power as well as other revolutionary technologies over the next 5 or 10 years, as already stated, there would be economic hardship for major companies. Not only would some of the oil companies go bankrupt or downsize, but so would some coal companies and power-generating industries, along with other companies such as drilling and mining concerns. However, many jobs lost in those industries would be gained in the new energy industries. The economy would get over it and would benefit in other ways already mentioned. Should we worry about a few industries over the life of the planet?

There is too much highly dangerous selfishness in corporate America. (Just examine the mortgage and banking crisis of 2008.) This is similar to the cigarette companies refusing to acknowledge the risks of smoking and purposely suppressing for years the known facts on the sickness and death that smoking causes. The fossil fuel and associated industries have done the same thing. The difference with fossil-fuel use is that we are not only talking about millions of people getting sick and some dying, but we are talking about the ruination of an entire planet and its inhabitants—a ruination that so many corporations are willing to risk for more profits. The corporate indifference to science and the health of the planet and its occupants has been really ugly.

The move to PVs, wind, ocean currents and other forms of innovative energy will be slower than need be as long as every dollar is squeezed from the fossil-fuel economy in house boilers and water heaters that light oil or natural gas on fire for heat, in stoves that light up natural gas, and in internal-combustion engines that light gasoline or diesel on fire for propulsion. The same scenario goes for burning coal for electricity or using bio-mass and ethanol. Even burning wood in a fireplace has become a serious air-pollution problem in some places, including the U.S., because too many people are doing it.

Do you think we went to the Middle East in 1991 to save Kuwait? If they grew potatoes over there instead of having oil underground, we would have stayed home. Was the 2003 invasion of Iraq really to stop terrorism and weapons of mass destruction? Was it because corporate America wants to open up new markets? Or was it to ensure a future oil supply? What did the gigantic corporate military-industrial complex have to say about it? Was the invasion for all of the above? Why did the U.S. support the shah of Iran? Why did the U.S. then support Iraq in its 1980 war against Iran? One thing is for sure, we are digging a deep hole of political unrest and future wars for a resource that we use incredibly wastefully, a resource that we should be frantically getting away from. And it's no news that we also support terrorism when we use gasoline or other oil products, as some of the money we pay to Middle Eastern countries finds its way back to terrorists.

Over 100,000 people died over oil in Desert Storm in 1991. Massive amounts of pollution were released from fighting the war and from burning oil fields to make sure Saddam Hussein didn't advance any further. If we are going to engage in something as radical as Desert Storm, to protect something as important as oil, then let's not be so incredibly perverse. We ended the war and came home, and over the following 15 years the order of the day was minimal conservation and maximum use of oil as the age of the SUV, the larger house, and extreme levels of consumption took off while our population exploded from record immigration. It's as though the American society closed its eyes and said, "Get me more oil, and I don't care what you have to do to get it. And I don't care what it does to my children's health or the health of the planet. I'm using as much as I want, and I don't care how many times we have to go to war over oil or how many people die." Did we ever stop to think of how much energy is used to fight a war (in terms of oil, gasoline, diesel, coal and resources of all kinds), and then how much additional matter and energy is used to rebuild everything just destroyed, not to mention the added pollution and overall environmental damage that all of this causes? The environmental damage (land and sea) from wars is extreme, not only from animals being killed, but also from trees and other plant life being killed, leading to habitat destruction and species extinction, plus all kinds of air and water pollution being released. A lot of greenhouse gas comes from fighting wars. Bombs, bullets, missiles and all kinds of weapons are designed to kill as many living things as possible, plants and animals, and to destroy as much material as possible, with no regard for pollution and its lingering effects. And we continue the process.

Look at our cities. Millions of lights are left on 24/7, from those in offices to neon advertisements. Times Square is an example of this madness. We could dramatically cut energy consumption, possibly by one-half, along with its associated ills, overnight, with a stroke of the congress pen and then a flick of an off switch. This new law: all lights get shut off at the close of business (Japan has started similar programs) except those linked to crime reduction or those running on renewables like solar. Ever fly over the U.S. and look down? Shut off every other street light on highways and roads to reduce electricity use, then lower the speed limit to 55 mph for safety and to decrease gas consumption. No more SUV's or any car that gets less than 60 mpg. Shut off the fax machine and computer, and get rid of standby power on all appliances, televisions and the like. Increase efficiency on everything.

Do we really need lawn blowers and edgers? Do we really need to mow our lawns every week? Lawn and garden equipment (blowers, edgers, mowers etc.) is responsible for at least 5 percent of U.S. air pollution from mobile sources. A gallon of gas burned in a mower can be the equivalent of a car driving 100 to 300 miles (depending on the type of car and its condition). All terrain vehicles, snow mobiles and jet skis pollute in a similar manner. All these devices may not produce the volume of pollutants from other sources, but they burn dirty and concentrate pollution near the source, which is why the typical suburban backyard can be hazardous to your health. Do we need all the gadgets we use, from hair blowers to electric knives and electric toothbrushes? The list of wasteful and damaging practices is long, and yet there would be little or no change to our lifestyles if we stopped or altered many of the things that we do. Some good things have been done. Cars and garden equipment are more efficient and burn much cleaner than they did 30 years ago, and many other improvements have been made, but as we increase the population, drive more and become more wasteful overall, good things are being canceled out by bad things.

The U.S. government could do so much more with new laws and programs. So far, they simply will not seriously pick up their pens, while the corporations write and direct the show. For example, why were the car manufactures allowed to follow much lower corporate average fuel economy (CAFE) standards for light trucks (SUVs, pickups, minivans) than could have been achieved? The standards for these vehicles were set at 20.7 mpg, while cars were set at 27.5 mpg (neither category met the standards). The petroleum and automobile industries lobbied Congress so intensely that the Clean Air Act Amendments of 1990 were not what they could have been, and higher standards over the years were simply not seriously considered by the Department of Transportation.

In May 2007 the U.S. Senate Commerce Committee passed a voice vote to raise automotive fuel standards to 35 mpg by 2020. The bill is not finalized as of this writing, and some senators are complaining that it is not fair to the automotive industry—of course. As would be expected, the automotive industry plans to fight the bill. In any event 2020 is way too late in the game, and 35 mpg is about half the mileage we should be averaging right now. In reality the senators should be discussing ways to virtually eliminate the need for gasoline by 2020.

Here's a sobering comparison: a GM Hummer vs. the Toyota Prius hybrid. The large Hummer burns approximately five times the gas for the same distance traveled and can emit more than five times the pollution. Let's drive down to the supermarket in a Hummer for a quart of milk, shall we?

If we look at over 4,000 people, including children, killed in 2004 in SUV rollovers and SUV/minivan small car impacts—with similar statistics since—and other large vehicle events such as reduced visibility that SUVs and minivans cause to other drivers, and SUV back-up incidents in which children are run over, we have another sad statement about how our culture operates. And in addition to the deaths, let's not forget the many serious injuries incurred from all these events. The parent who says he or she wants an SUV to keep his or her child safe is off the mark. Burning the extra gasoline is potentially ruining that child's health and environmental future, and when the SUV rolls over, that child may die—and when it hits a small car, someone else's child dies. People will always be killed in car accidents; this can't be stopped, but it can be lessened if we drive slower, separate trucks into specific lanes, and are more careful. Small cars can also be built much stronger with a minimal increase in cost, and the result would be still not as expensive as some SUVs. Get the SUV's and minivans off the road and the self-induced problems go away.

Hybrid SUVs are here, but the development logic behind them is tainted or nonexistent. The marketing is profit driven and directed to an ignorant public. Ford's new SUV, the 2008 Escape hybrid, claims an EPA estimated 34 mpg in the city, and 30 mpg on the highway. Hybrid SUVs in general are boosting their mileage and therefore a lot of them will likely sell. Consequently, we go backwards because buying these hybrids stops many people from buying the hybrid car that gets 55 or 60 mpg, and slows the advancement and purchases of the 100 mpg-plus hybrid soon to be on the market. Even if hybrid SUVs eventually go plug-in, they still won't get as much mileage as a car. Unfortunately, too, hybrid luxury cars are being produced that get 35 mpg or less. Some conventional cars now get more than 40 mpg on the highway and 35-38 mpg in the city. Why would anyone consider buying a hybrid that didn't get the technology's maximum mileage? This defeats the whole purpose of going to hybrids in the first place.

The reality is that the car companies still market to vanity, ego, status, desire and people's beginning acceptance of the newer environmental movement. The green revolution is here. But unless we are getting the maximum mileage available, then regardless of what the car makes us look or feel like, we are not green. We've been had and marketed to once again. End result: we still burn about twice the gasoline that we should; solar and hydrogen are further delayed; pollution problems continue, and global warming accelerates.

We are attempting to solve our energy problems in the United States but we are doing it at a corrupt-snail's pace, maybe not fast enough to head off potential economic and environmental disasters. Even if we went totally to solar power and other clean sources of energy in the next 10 years (and meaningful change is unlikely in 25 years unless someone in charge gets a grip), we would still have other very serious problems that we are not facing.

We have built the world's largest economy on energy from fossil fuels that are running out and simultaneously killing us off, and we continue to expand and grow economies based on these same fossil fuels. We harbor the assumption that the earth's ecosystems have an unlimited ability to absorb the damage that we do. We also ignore many other problems that come with growth and expansion, problems that arise directly from overpopulation.

6
OVERPOPULATION

Ten thousand years ago—as best as can be figured—there were two to five million people on earth. As of mid-2008, there are about 6.7 billion people inhabiting earth, a planet that is approximately 4.6 billion years old. The earth took a long time to evolve to its present life-supporting state, and during the last 600 million years or so, life started to seriously evolve into what we know it as today. Before that time, known as the Precambrian, life was very different than it is now. Many life forms lived in the oceans, while on land nothing substantial could exist because the ozone layer wasn't fully in place yet to protect living things from deadly ultraviolet rays. It's been a long road for plants and animals.

We humans are a blink in time, a grain of sand on a beach. The analogy of a 350 page paperback has been used many times. If the book's words represented earth's lifetime, then human existence might be represented by the last few words in the book. We haven't been here long at all. Many other species have been here much longer, such as sharks (about 400 million years) and crocodiles (about 60 million years). Our primate ancestors only began to take shape approximately three million years ago. The evidence is still being sorted out, with new discoveries happening as I write, but it seems Homo sapiens as we know ourselves today finally stood out about 125,000 years ago. We were still primitive back then, but our height, weight and brain size were in place so that we were similar to what humans are today. As far as organized development of a culture goes, we can pick whichever one we like: the Egyptians, the Greeks, or the Romans, collectively spanning a time frame of 3,500 to 2,000 years ago, are examples. But what really matters is modern humans of today, because no matter how advanced ancient civilizations were, they could not damage the world as we do today. Their damage was localized.

Most of the environmental damage humans have done to the entire planet has taken place during the past 150 years. This damage has escalated over the past 50 years as modern technology has advanced dramatically, population has grown and global consumption has increased. This 150 year assault on the earth has caused more environmental change than nature might in thousands of years. In the case of carbon dioxide in the atmosphere, we have increased this greenhouse gas to levels not seen in nature in 600,000 years or more. We have dramatically increased other greenhouse gases as well.

A five degree Celsius (C) average change in temperature worldwide can mean the beginning or the end of an ice age. Today we may cause from a 1.5 to as much as a 6 degree C (10.8 degrees Fahrenheit) change over a mere 100 years. Our existence is nothing when compared to the approximately 3.8 billion years of life evolution, from the first bacteria, yet our effect on the planet stands out significantly.

Figure # 1, World Population, tells part of the story. (All figures are near the middle of the book, following this page.) The data go back to about 1,000 years before the present, and project into the future. Up to about 1000 A.D. world population grew to approximately 500 million people. Then for the next 800 years or so, population grew gradually to approximately one billion by about the year 1830, and it might have gone a bit higher but would have eventually leveled off or declined, depending on what took place based on nature's rules, but something extraordinary kicked in. There was a major change in the way human beings lived on earth. That change was the famous industrial revolution, which included manufacturing, agriculture and medicine. We started to become modernized. Before this we were still relatively primitive, and therefore the same things that kept every species' population in check, such as disease and the ability to obtain food and water, were also obstacles to dramatic human population growth.

Other animals—everything around us that crawls, walks, swims or flies—exist because there is a enough food, cover and space in a given time frame (we will include water with food). Remember, there are about 1.8 million known species of plants and animals on the planet, and humans represent only one of these species. All species, both plants and animals, need food, as well as cover (plants less than animals), meaning a safe place to live and be protected from the elements; even a deer or bee will die from exposure to weather. We also need space. Food and cover are really self explanatory, but space is not.

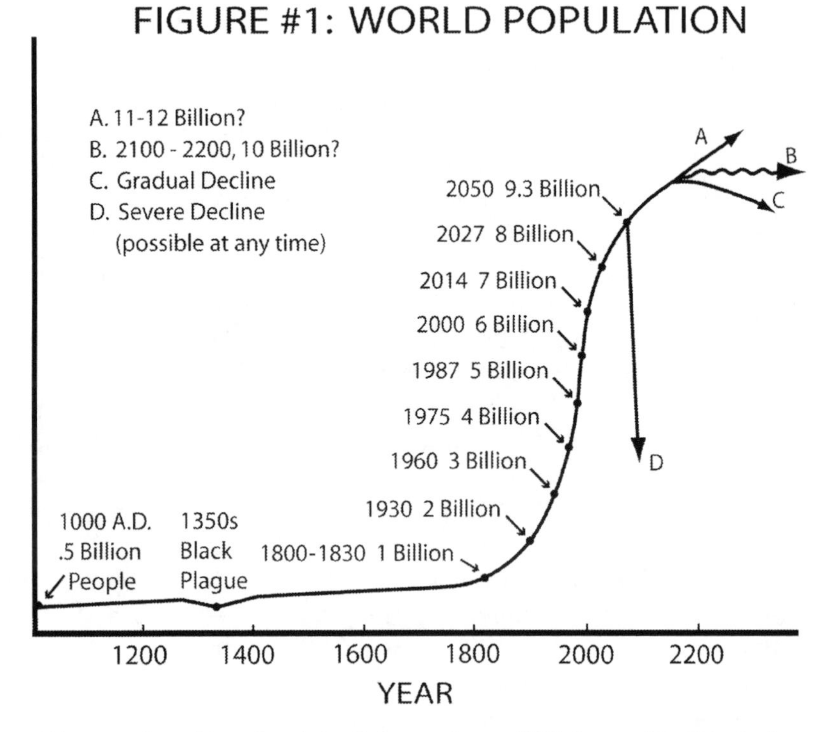

(Diagram adapted from Population Reference Bureau (PRB) data, and Graphics Bank: Population Basics: World population Growth Through History, and World Population Growth in Billions, 2006)

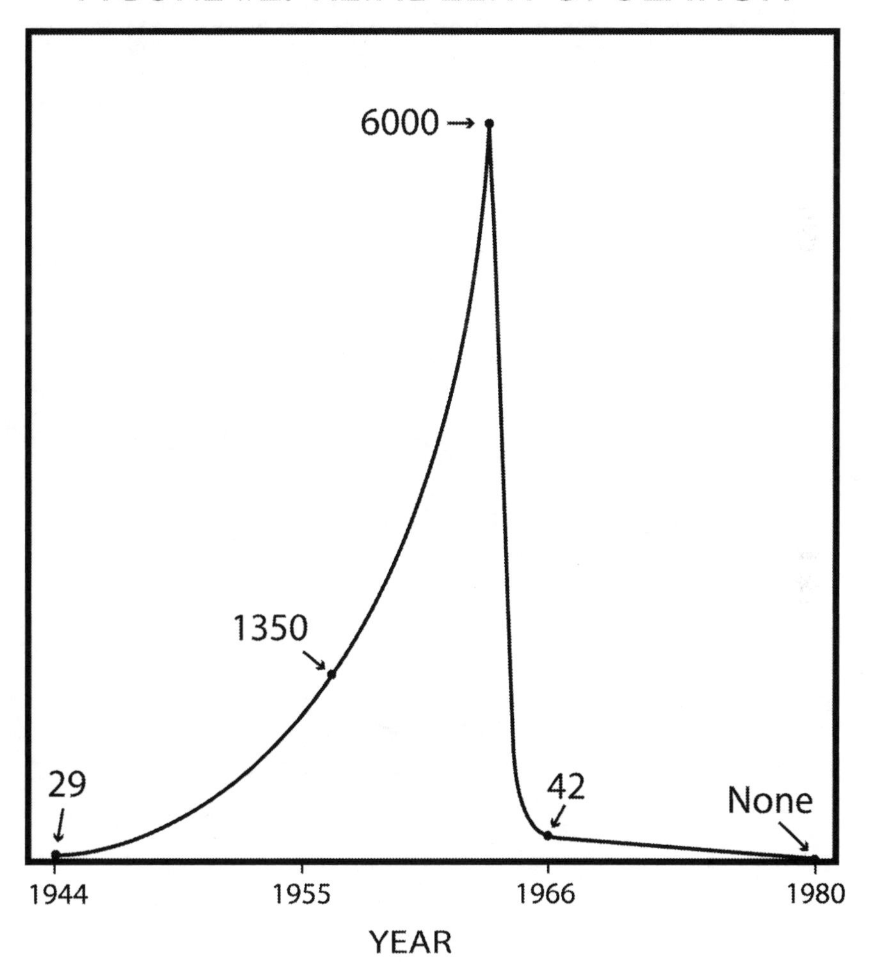
FIGURE #2: REINDEER POPULATION
6000
1350
29
42
None
1944
1955
1966
1980
YEAR

FIGURE #3: POPULATION

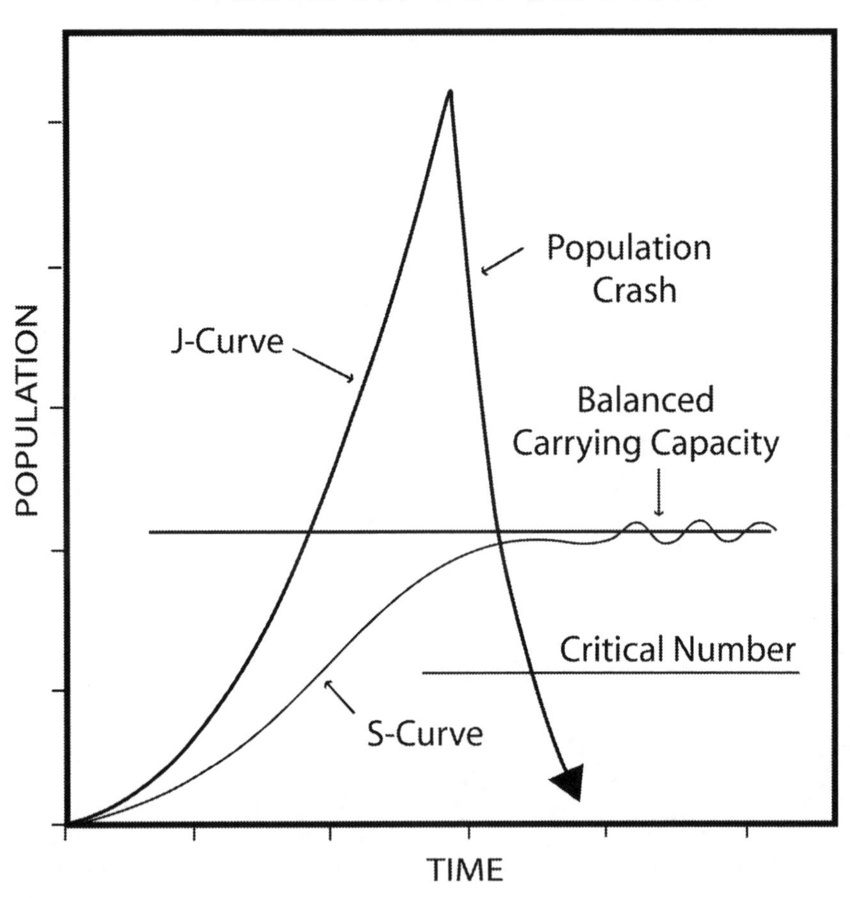
Population
Crash
J-Curve
Balanced
Carrying Capacity
Critical Number
S-Curve
POPULATION
TIME

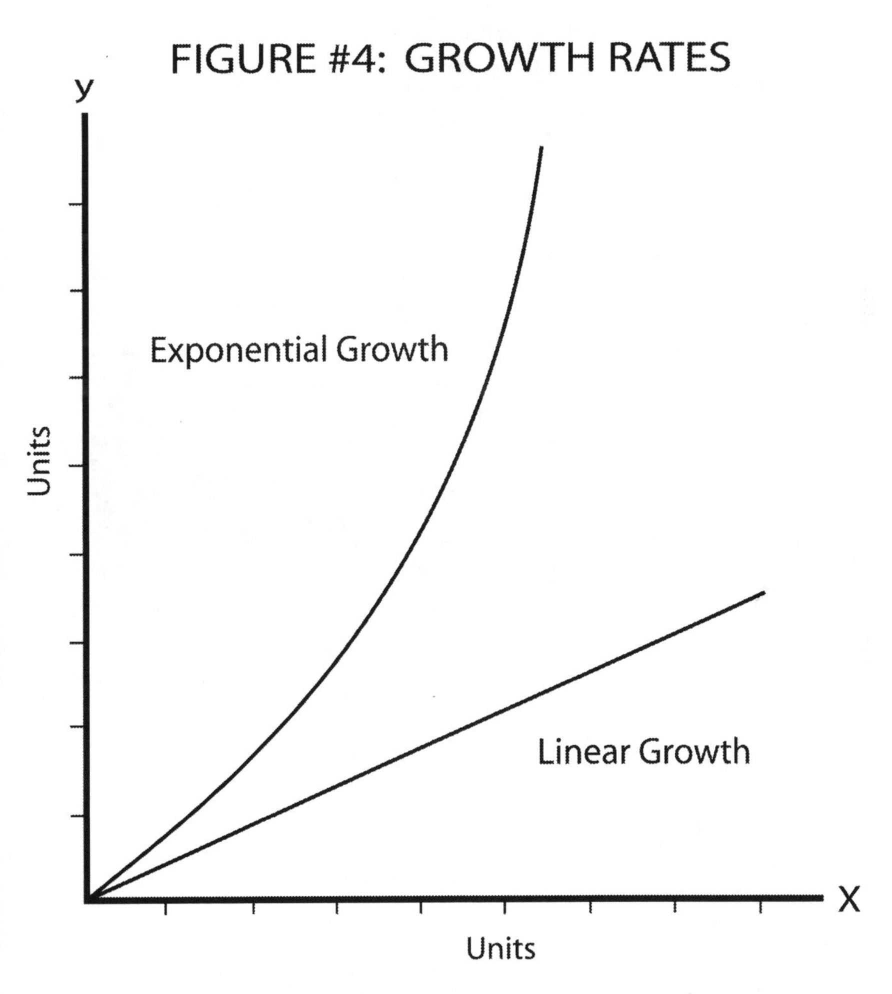

Exponential growth represents a rapid and extreme increase.
Linear growth represents slower, uniform growth.
Units can be anything, for example, time on the x-axis and human population on the y-axis.

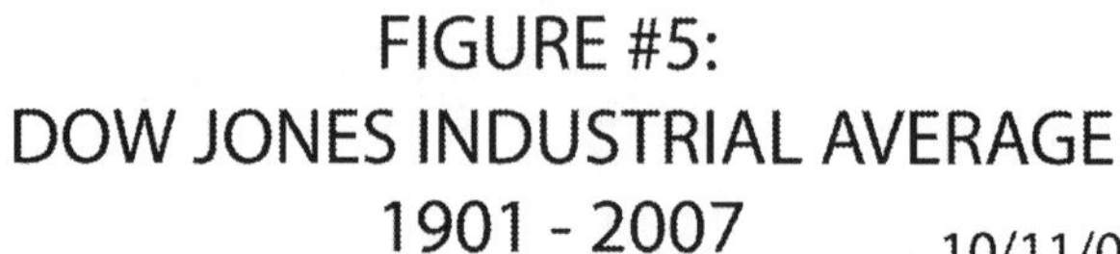
FIGURE #5:
DOW JONES INDUSTRIAL AVERAGE
1901 - 2007

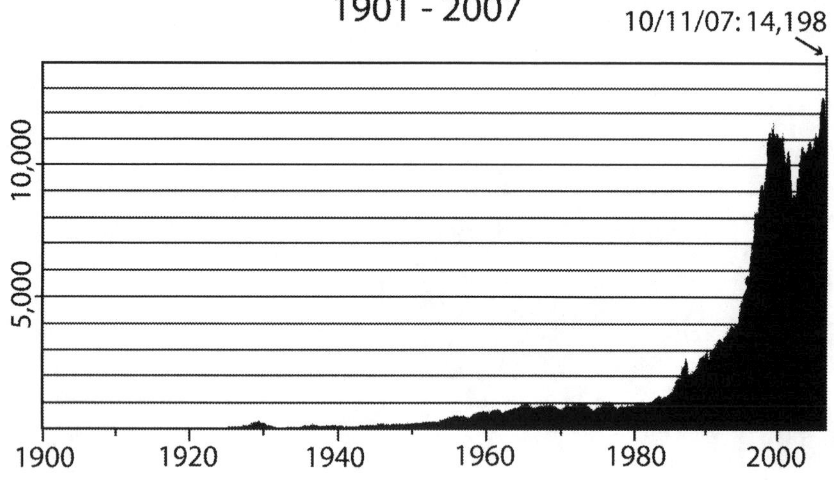
10/11/07: 14,198
10,000
5,000
1900
1920
1940
1960
1980
2000

FIGURE #6: CONCENTRATIONS OF GREENHOUSE GASES FROM 0 TO 2005

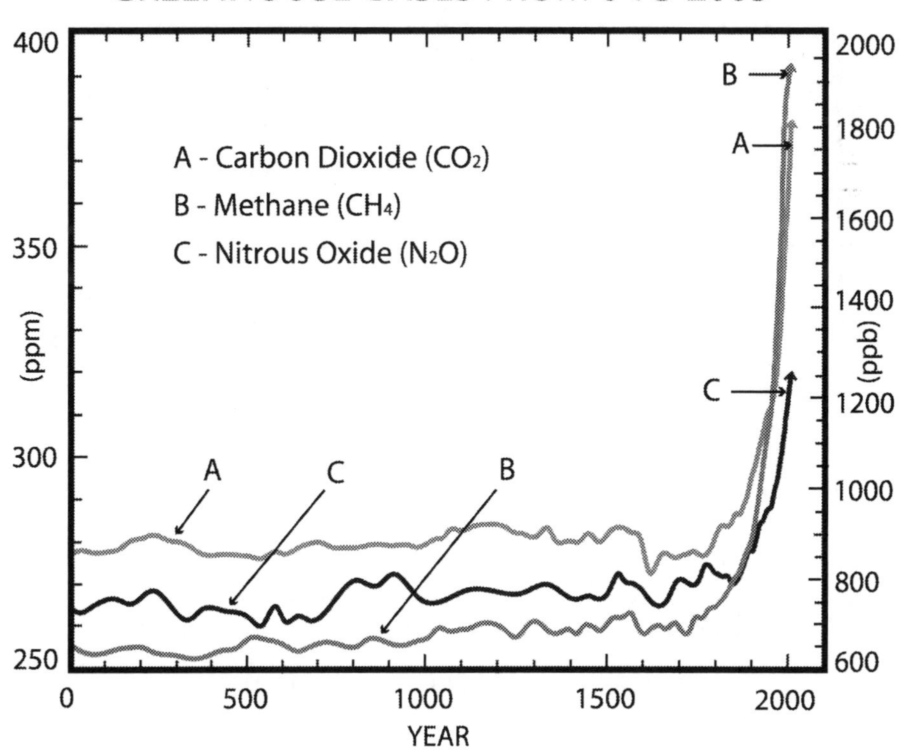

CO_2 (ppm), N_2O (ppb)
CH_4 (ppb)

(Source: Intergovernmental Panel on Climate Change (IPCC), Climate Change 2007: The Physical Science Basis, Climate Change 2007: Synthesis Report)

FIGURE #7: TEMPERATURE CHANGE (PAST 1000 YEARS) NORTHERN HEMISPHERE

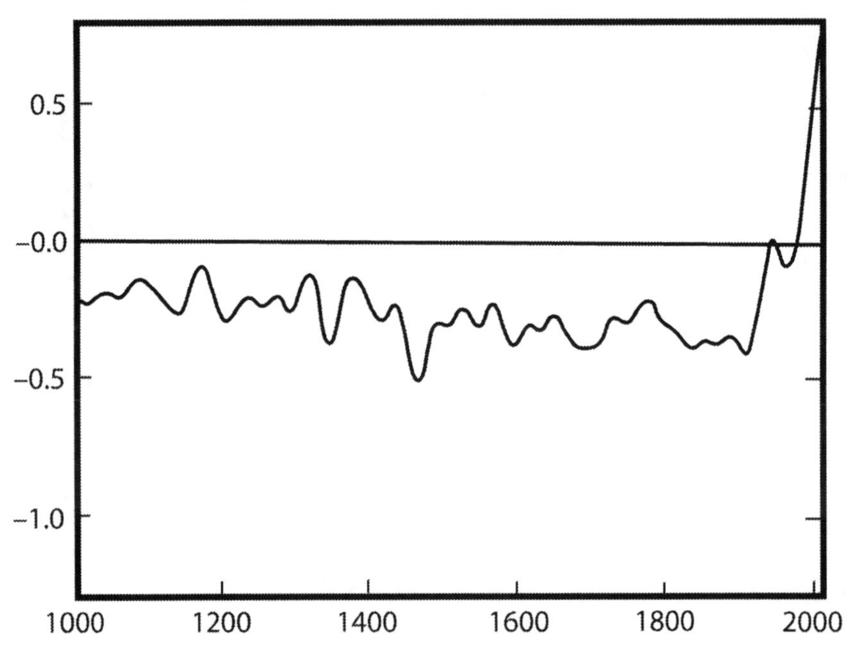

Departures in temperature (°C) from the 1961 - 1990 average

(Source: Intergovernmental Panel on Climate Change (IPCC), Climate Change 2007: The Physical Science Basis, Climate Change 2007: Synthesis Report)

Some years ago a population density study was done on rats. It's not the only study done on the subject, but it was a good one. In the first part of the experiment a small group of rats were put into an outdoor enclosure of one-quarter of an acre. At the end of 27 months the rat population had increased to about 150 animals and then stabilized. The population should have gone higher, given 150 healthy rats, but did not.

The reason was determined to be that infant mortality was abnormally high due to stress from forced social interaction arising from the unnatural confinement, causing maternal behavior disruptions among the adults.

The next part of the experiment was taken indoors, where another small group of rats were put in a room, about 10 by 14 feet, that was partitioned into four quadrants with openings in all the partitions so that the rats could freely move about and intermingle. The rats had all the food and water they needed, as well as comfort and protection from the elements. The food was placed in one of the quadrants where all of the rats met at feeding time for a free lunch. During a 16 month period, the rats did what all good species do: they had families. With plenty of food and cover there was nothing else to do, but with too much of a good thing something started to happen.

As their population grew, the room got very crowded, because space was the only thing the researchers wanted limited, obviously by the four walls. In time the rats became very aggressive, started to fight, deaths occurred, there was cannibalism: some of the young were killed and occasionally eaten. Reproductive organ failures took place in some females, and some females even died while giving birth. After giving birth, many females did not fully perform their maternal duties; they stopped building nests, seldom nursed and even abandoned their young. The male rats displayed berserk behavior, attacking females and the young, engaging in cannibalism and sexual deviation. Other males became more passive. Some rats withdrew from the group and would only come out to eat, drink and roam the compound when the others were asleep.

What the study showed was that lack of space—that is, increased population density—creates psychological stress that leads to severe social pathology. Indeed when habitats are destroyed—which decreases space—animals do die off because unnatural things start to happen. Of course in nature, rather than a controlled experiment, the diminishing of space also causes a reduction in food and cover, which eventually leads to additional problems, such as death.

When domesticated animals raised for food are penned up and severely crowded for months on end, many do suffer from the lack of doing what comes naturally, that is moving around and engaging in behaviors such as foraging or just simple exercise and grooming themselves. In Venezuelan rain forests, monkeys—many other animals too—have been crowded into smaller and smaller spaces as their habitat is destroyed and fragmented by human activities. Some have responded to the stress by becoming more aggressive, and some mothers have also stopped nurturing their young. Habitat destruction is a worldwide crisis for plants and animals, not only in affluent societies but also in poor areas where increasing human population crushes everything in its path.

In prison humans are forced into crowded conditions, as were the rats. This is one reason why prisoners are periodically allowed out to exercise. If they weren't allowed out of their cells, they too would go crazy, but many, if not most, still suffer from the overall tight and crowded conditions, suffering sometimes expressed as violent and irrational behavior (not that all prison problems arise from overcrowding). Humans may be more adaptable to stress than other animals, and certainly we don't behave quite like the rats, but we definitely have our limits. We are social creatures, so we sometimes crave the company of lots of people, but we can get away from them when the crowding becomes overwhelming simply by going home. Even if that home is only a small apartment, it does have almost everything we need. But often the effects of being around too many people, too much traffic, long lines and the pressure to rush and to do more with less time—much of which is brought on by overcrowding—does catch up with us. It may be that as the world gets more and more crowded, stress related illness will become even more common than it is today—and it is already very common. Road rage is a modern term and may be a manifestation of the problems revealed by the rat study.

Stress is a major problem in the United States as well as in other parts of the world. Many times the effects of stress are not recognized and can be delayed. Heart disease can be caused directly by stress, and damage to the heart and arteries as well as high blood pressure have been the result of plain old stress over too long a period of time. Even sudden cardiac death can be the end result of a stress overload. The human animal can handle some degree of acute stress, the type that comes and goes very quickly, as in the fight-flight mechanism that all animals have built into them. When a dangerous event takes place—

typically a predator-prey relationship—adrenaline and cortisol flush into the body, preparing us for a fight or a flight (running away), but the event ends quickly, and the body chemistry goes back to normal—if you haven't been eaten. These mechanisms help ensure survival. What nature didn't plan for in the human animal, or any animal, is the ability to deal with the chronic stress that we are exposed to in our unnatural, modern, hurry-up world. This stress can cause the body to be perpetually out of chemical hormonal balance and over time can cause serious problems, including depression, insomnia, anxiety and sickness.

As it becomes more and more difficult to maintain the middleclass lifestyle, especially when money problems set in, then not only does the lack of space create pressure, so does the constant grind of trying to get more money to buy the food, cover, and space needed to survive. Throw in the fact that more and more of the United States is becoming city like and crowded, and we end up with a public that is generally overworked and burnt out while our quality of life is diminished. Ever hear the saying the best things in life are free? Getting that free and necessary space will become difficult as overcrowding reduces it and the richer appropriate part of it and drive up the price. This has already begun.

There are those that accept modern stress as part of life, but I would counter with the idea that the lives that many of us are leading are not healthy, and that humans cannot cope with the amount of stress—and bad diet—that our modern society puts on us. If not for technology that gives some of us our own unique space to hide in, with food and cover to boot, we too would exhibit some very strange behavior, exposed to the same situation the rats experienced. The problems resulting from overcrowding catch up with other species faster than with modern humans, who have technologies to temporarily solve the problems, and therefore defer them to a later date, but the problems will eventually catch up with us much more than they already have.

Consider a wolf pack roaming the West, on the lookout for food, elk being a favorite menu choice from which to gain energy. The pack can be 10 or 15 strong, sometimes more, and they need large areas to roam. (The arctic wolf can range 1,000 square miles.) Why? A lot of territory is needed for the predator/prey relationship. In other words, wolves need room to find their food supply, elk (they eat other things, too) and elk also need lots of space to roam for the same reasons, like finding things to graze on. In nature the whole ball game is who eats whom, or what—there are no refrigerators—and we are part of this game, like it or not. The pack also needs water and needs to find cover to sleep and for protection from the elements. Minimize this space, and wolf packs are pushed into each other's territory, creating intraspecific (within the same species) and interspecific (between different species) competition that otherwise would not have existed. A dangerous situation will then develop, threatening the long-term survival of at least one of the packs, possibly the entire species.

Watch the animals in a zoo. Some of them are extremely stressed from space being limited. I was once at a very large, famous U.S. zoo where the polar bears and wolves looked like they were going out of their minds. The bears, animals that normally range tens of miles in a day, were living in a very small area and were swimming in tight, 20-foot circles and would not stop. The wolves were rapidly pacing back and forth behind a caged-in area with a look in their eyes that was very sad. They had all the food and cover they needed, but not the space. The handlers admitted the animals were under great stress from confinement.

Food, cover and space are the ingredients that allow all species to survive. Unfortunately, though, nature has additional rules for the game. Disease gets thrown in as an added feature to balance the act. Population does not J-curve for too long. J-curves only happen for short times and only due to unusual circumstances. Look at Figure #1 again. It's a picture of a J-curve (a curve similar to the letter J) that represents extreme population growth. This is a picture of a population rising contrary to environmental resistance and therefore suggests that food, cover and space are limitless for that particular species. This would also then suggest that the laws of physics are false. For any population to grow indefinitely or for any very extended time period, food, cover and space—and therefore matter and energy—must exist in an endless or unrealistically large supply on a planet with definitive boundaries. A J-curve for an extremely dominant species also implies that many other species will perish so that one species may grow its population past where it belongs.

If ant populations grew exponentially, with none of the restrictions that nature and physics keep in place, then soon ants would be crawling over almost the entire planet's land mass. Most other species would be killed off, including us, as the ants dominated and appropriated everything. But then the ants would rapidly die off when eventually there was nothing left for them to survive on, because matter and energy are not limitless. This elementary scenario is exactly what humans are engaging in. The difference between us and them is that they can't overpopulate for extensive time periods, but we can. What is so worrisome is that so many of us don't understand that our J-curve too, is only temporary. Further, we shouldn't let the leveling off at the top of the J-curve at point B, in Figure #1, fool us.

A classic example of what happens to a species that J-curves its population is the 1944 case of 29 reindeer being introduced to St. Matthew Island in the Bering Sea between Alaska and Russia. There were no natural predators, and there was plenty of food to graze on. Over a 19 year period, by about 1963, the population of reindeer grew from 29 to 6,000, at which point the 128-square-mile island became very small. By 1966, just three years later, there were only 42 animals left (Figure #2) after mass death from starvation and exposure took over and a dramatic population crash ensued. By 1980 the reindeer had vanished. Having exceeded carrying capacity, the reindeer destroyed their life support systems and caused serious damage to the island's ecosystem from overgrazing. This kind of assault will cause survival problems for other species of plants and animals that rely on ecological stability; such is the connection between all species.

Another event took place on a 45-mile-long island called Isle Royal, in Lake Superior, where moose found their way to the island, but this time wolves followed later on. A rise in the moose population was tempered by the wolves as predators, as balance was injected into the system. The two species followed each other up and down in population as nature strove for the end result of an S-curve (Figure #3) in both species. In 1980 humans unknowingly cut in by introducing a canine virus that killed many of the wolves. The moose population temporarily J-curved, then crashed. The process was helped along with a harsh winter and an infestation of ticks.

With an S-curve, population levels off and flattens out, making the entire curve look like an elongated letter S, more or less. All the same, nature doesn't easily go from a J-curve to an S-curve, because the population that has J-curved has likely already passed its carrying capacity, so there is a crash in population in order to bring things back into balance. From the beginning, never getting to a J-curve, but following a much more gradual S-curve is what nature would prefer—but doesn't always get; nature is also chaotic yet maintains a balance within this context—because it represents slow, sustainable rise of the particular species to the top of the S-curve at which point the population levels off. This leveling off to a relatively steady state is the species' carrying capacity, which is the population level that an ecosystem can support under given conditions of food supply, cover and space (including predator/prey relationships, etc.), which are subject to change over time. When one or more of these things are limited, then the animals become weakened, some may starve, and disease has a better chance of bringing down the population. A new, lower carrying capacity may then be established, unless the population has dropped down past the critical number below which it can't come back. That critical number (Figure #3) is simply defined as too few of a species to re-establish itself.

The scary thing about J-curves is that a dramatic population crash is ultimately unavoidable, and the critical-number point can then be rapidly and severely breeched. The species in question then may not come back, or the population may only be severely decreased. Technology can prolong how long the J-curve and the population increase continues, but this only delays the arrival of the day when the inevitable happens and everything falls apart. Take another look at the human J-curve in Figure #1, and then look at Figure #3 again.

We have an enormous problem as humans. We are in the midst of a gigantic J-curve in our own population, and we have no predators. We have cured many diseases and diminished bacterial infections that once limited our population growth, while technology has allowed us to overcome other natural obstacles to our growth by draining the earth of its resources, ruining the earth's other species and ecosystems, and polluting the planet. We are artificially driving up our species numbers. The leveling off of the human J-curve into an S-curve (Figure #1, point B) is a deception, because it assumes that at this point we have reached our population's carrying capacity.

The coming S-curve is only a calculation, an assumption, as to where we think our carrying capacity might be, that is, where we will level off our population growth by our own choice because of changing our cultural and economic situations over time. However, this leveling off is not nature's choice because we have temporarily gained the ability to ignore nature. This artificial, unnatural and dramatically steepened S-curve that we have created on paper because of changes in our cultural and economic norms does not correlate with the natural world. This false S-curve is very dangerous and no more represents our carrying capacity than would be the wolves' collective decision to buy refrigerators to store food, in order to raise their own carrying capacity.

If we were still primitive, then we would be in the same boat as other animals and there would be limits we could not overcome. There would be no way for us to increase our population very much past the carrying capacity that any ecosystem during any given time established for us. Because we are modernized and do things that are damaging to the natural world in order to maintain our lifestyles, carrying capacity has to be looked at in a different way. We must look at our consumption levels and the damage that this consumption does to the planet. In the human time frame, much of the damage is permanent, and so we are constantly diminishing the planet's ability to support us while at the same time we continually increase our numbers and consumption levels. Also keep in mind that in nature, all matter and nutrients are recycled—that is, reused again and again. All organic matter reassimilates when animals and plants die, when animals defecate, urinate and spread seeds. A plant regrows because of death; life is constantly reborn as the sun gives the new plant the energy to grow. A molecule of nutrient or water that we ingest may have been in the body of a dinosaur or gladiator way back when. Naturally, everything keeps turning over in a clean and healthy manner. However, humans return almost nothing back to the earth in an easily useable condition. When oil gets taken out of the ground, it is returned as air or water pollution, or plastic or chemical waste. There are too many ways in which we pollute by extracting, manufacturing, consuming—and then putting back pollution and garbage. Some people say that the way we recycle things such as metals, plastic, paper, etc. is proof that we are paralleling nature, but our methods are not always in sync with nature. Our recycling still uses energy, pollutes and puts little back naturally, as already mentioned in the energy chapter. The main thing that our recycling does is reduce the amount of new resources that we take from the earth, but this is not the cure.

Our population numbers have gotten where they are because we have been destroying other animal and plant life and degrading the planet's life-support systems. In order to keep growing our population and economies, we are borrowing at an accelerating rate from the carrying capacity of every other species on the planet, plant and animal, and we are not paying back the loan. At the leveling-off point that we may or may not reach somewhere between 9 and 10 billion people, the drain on the planet may only change from an accelerated rate to a constant rate. However, if we reach a plateau in population but keep drawing more and more people into the consuming lifestyle, then the drain on earth will continue to accelerate to even higher levels as pollution continues to increase. This type of drain on ecosystems and resources will continue to result in extreme damage, making the larger human population far less sustainable than our population is as of 2007 - 2008.

What is the carrying capacity for our modern human world? It is very difficult to say, and making a definitive statement about an exact number would put any scientist on thin ice. However, I am confident that the earth would be unable even to support its current population of over 6.7 billion people if everyone were producing and consuming at the same rate as the U.S. is today. It is very doubtful that for any comfortable length of time the earth could support three billion people living like we do in the United States while the remaining population continues to consume, albeit at lower levels. This should be clear given the damage we have already done to the planet's atmosphere, oceans, ecosystems, and all life, with only about a billion people living in any way similar to the U.S. Negative Population Growth, an organization dedicated to population issues, has estimated that a world population of two to three billion people would be optimal if everyone lived in environmentally friendly ways. For all people to be well taken care of this is probably a safe estimate. Clearly if everyone lived more primitively, then the planet could support larger numbers, but as it stands, the way human societies are functioning, it is very likely that the 6.7 billion people that exist today are too many.

With conservation and technology we can minimize damage and support human life with dignity, but if we use these things as an excuse to increase population and consumption, rather than to create a fair balance between nations, then we will run into a point of diminishing returns. Ultimately, the lower our population is, the safer we will be, and being below carrying capacity is the safest place to be.

Population growth and consumption must be stopped and stabilized. With advanced technology and conservation there is a modern carrying capacity, one that factors in consumption levels, and what ever it turns out to be, it will require the population to consume and live in a rational manner, which would also require an environmentally educated, environmentally conscious worldwide society. At the moment, the United States—along with other parts of the world—is not yet fully environmentally educated or conscious. In the U.S. increasing population and consumption is encouraged, but instead, overpopulation issues and carrying capacity should be openly discussed at every government level and in every campaign debate.

Aside from the human-induced problems such as global warming, there are also natural changes that will take place. Events such as large volcanic eruptions, large earthquakes, tsunamis, hurricanes and tornadoes are all going to happen periodically. That's the way the earth works. Some events are more problematic than others, and hurricanes and tornadoes may be seriously intensified by global warming. The larger and denser the population in disaster-prone areas—and there may be new disaster-prone areas as global warming kicks in—the larger the loss of life and the economic loss will be, as was the case with Hurricane Katrina in August 2005, which turned out to be the United States' worst natural disaster. As population centers grow, so will the number of people in harm's way, while the likelihood of a hit also will increase. This concept is obvious, although some people in charge seem to be oblivious to it.

Federal and state governments were warned time and time again for many years that a New Orleans disaster was going to happen in part because of human induced changes to the Mississippi River and the wetlands that protected the coastline. Over the years newspaper articles, magazine articles and television documentaries pointed out the problems and risks in detail. The warnings included factors such as levees that were not up to the job and the simple statistical likelihood that a very large hurricane—global warming intensified or not—would eventually hit the area. I was lectured to some 25 years ago about the real possibility of a disastrous New Orleans event—the details of which were exactly what happened—by my first environmental science professor. Shortly after New Orleans was wiped out by Hurricane Katrina, President Bush made the statement: "No one could have predicted such a devastation."

The December 2004 tsunami that hit Indonesia showed clearly what happens to densely populated areas in such an event, especially when the population is so close to sea level. If the people were warned in time and a lot of lives saved (over 200,000 died), there would have then been several hundred thousand more homeless people. Where do we easily put those additional homeless people, and how does their continued living expense get paid for? After Katrina tens of thousands of people in the U.S. were homeless and displaced, which creates an extreme psychological burden for those people as well as an ongoing financial burden for everyone.

Another extreme event in a densely populated city, or over a large area encompassing more than one city, is possible. A global-warming-induced or exacerbated event—super hurricane, pandemic, food and water shortages—

or an everyday tsunami or terrorist attack (nuclear or chemical) could displace millions of people, and even kill millions in the U.S. A large event will cause emergency services and hospitals to be so overwhelmed by so many homeless, injured or sick people that a lot of people will be left to die.

For the next 20 years or so we have entered a period of natural increase in hurricane activity. Add in the likelihood that global warming will make the hurricanes worse, and we may have a genuine problem. We did break records for hurricanes in 2005, and hurricane intensity in the North Atlantic has intensified over the past 30 years. It is not impossible that the perfect storm on the perfect track could run a wall of water over parts of Manhattan and Long Island and devastatingly affect millions of people. It is also quite conceivable that multiple events could take place at the same time in the United States or the world. In the U.S. high population density caused by overpopulation and how this relates to creating dangerous conditions when disaster strikes is quite unfortunately not viewed as a problem—nor is our overpopulation viewed as an issue in terms of how it contributes to the world's environmental problems through excess resource use and pollution—especially carbon dioxide emissions. We keep growing our population in order to expand the economy with no regard for anything other than corporate profits and government tax revenues.

A convergence of disaster-related issues is going on right now in the world, as well as in the United States. Our population is exploding to unprecedented and un-supportable levels, at the very time that global warming is starting to become a threat to all species' carrying capacities. Even if we did not increase the population any further, global warming still would threaten food supplies, water supplies and the very existence of coastal areas where about half of the U.S. population lives. As our population and its density increase, worsening global warming as well as other environmental problems, we will need more and more food, water, energy and manufactured goods, as well as space. We are increasing our demand for these things by the minute, yet these are the life-support mechanisms that global-warming-related issues will take away. In short, we are driving down the earth's carrying capacity for us while we drive up our numbers and demands on the earth. We are setting ourselves up for a horrendous fall from which we will not get back up easily, if at all!

The earth's other species continue because of their fitness, sometimes called Darwinian fitness (after Charles Darwin, 1809-1882, the naturalist who originated the theory of evolution and natural selection), often simply called survival of the fittest. However, the definition is not a simple one. The term "fitness" is used to describe a species ability to mate and reproduce and pass on strong genes in order to survive. This feature includes the ability to adapt to selective pressures such as natural climate change, drought, food shortage, or disease, as well as the ability to be involved in a predator/prey relationship. Animals need to produce high-quality offspring to survive. A male wolf with weak legs does not normally mate with a healthy female. This mating could be a disaster for the pack, as wolves might eventually be born with weak legs and so would not be able to hunt adequately.

Human beings no longer survive according to pure Darwinian fitness. We run on a modified version that could be called human fitness, which incorporates economics. Those with the most money essentially can buy what they need. We do pass on genes, but if genes are unhealthy, modern medicine can many times solve the problem, if one has the money to buy the doctors and medicine needed, outright or through health insurance. Someone who is handicapped can still live relatively well. Even terminally ill people sometimes can prolong their lives with enough money. Of course there are limitations, but to a great degree, money is the "gene pool" to be passed on in the human world.

Most modern humans do not need to hunt for food, tend crops or protect against attacks from neighboring tribes. We don't have to build a cabin or shoot and gut a deer in order to eat. Our society is structured to maintain itself with infrastructure and technology such as refrigerators, air conditioners, modern homes and medicine, supermarkets and other conglomerates that take care of our needs with machines, computers and warehouses full of what we need every day. A complex social structure of government and services such as police does the rest. For people with very little money, life can be difficult, and for people with a lot of money, life is clearly better (leaving out emotional problems, etc.). Money and economics are the Darwinian fitness that humans use as measures of success and strength.

Human fitness might be called pseudo-Darwinian fitness, or sometimes it is called social Darwinism, because although we do compete with each other for food, cover and space, we also compete with each other for desires and relationships, for money and business, and for land and resources. Wars take place over all kinds of material wants that sometimes have little to do with need but more to do with greed. Since the beginning of human existence, year after year we have killed millions of people in wars, invasions and takeovers, from the Romans slaughtering Christians to the Europeans taking the American Indians' land, to World War I and then World War II, with many smaller wars in between. In this sense—and even more so in modern times—the animal with the strongest genes and biggest claws and teeth is really the one with the biggest and best weapons, and money buys the weapons.

The earth doesn't recognize racial, cultural, religious or economic concepts. The earth doesn't care about us as a species, but it does give us what we need to survive within the realm of Darwinian fitness and maybe just a little extra. It does this with the same ruling given to all other species. That is, if we use up the earth's resources and therefore its ability to support us, or if we damage the biosphere enough, then we will go the way of the reindeer on St. Matthew Island, and our population will crash. What might bring us down first is speculative, but disease due to climate change and overpopulation is a high probability. Water shortages caused by radical changes in the hydrologic (water) cycle are also a threat.

In a worst case scenario, another real threat is massive food shortages around the world arising from severe climate change affecting crop yields and pollinators. This could mean starvation and diseases related to starvation causing deaths on an unimaginable scale, even in the United States, where tens of millions could die. When crops fail, that means milk, eggs, beef, chicken, pork, bread, cereals—and almost everything else on the supermarket shelves—are no longer plentiful. The shelves may even be empty. Fruits and vegetables don't like climate change either, nor do the animals, such as bees, that pollinate them. Our damage to the environment threatens not only the domestic food supply that we grow but also the naturally growing foods. Marine life too, would suffer from severe climate change, and today we are already seriously affecting the ocean's ability to supply us with food. And just like people, animals die directly from climate change. Cattle and people have died in the U.S. from record heat. This has been happening all over the world. Not only do crop yields decrease in higher temperatures, but even milk yields from dairy cows decrease.

One thing is certain. We are the most dominant species, ruling inhumanely and irresponsibly over all other species, and we are increasing our population and consumption levels as though food, cover and space were limitless, as though energy and matter don't matter. This reckless behavior is dangerous not only for the obvious reasons of direct destruction to the environment but also for the social implications it imposes on us—extreme competition between humans leading to world war, or in any given country, a revolution and the collapse of society—that could bring about our demise.

When any species overpopulates, life becomes less valuable. But nature doesn't allow for cheap life, so nature takes care of overpopulation in ways that are deemed cruel by us. But are they really cruel? The obvious way that nature balances the act is simply by letting animals or plants die when food, cover or space becomes limited. Animals do sometimes actually lie down and die from hunger or lack of water. Animals also sometimes do necessary things that are quite different from what we might do when problems arise. When a bird drops one of its two offspring out of the nest to the ground far below in order to kill it, so that the other offspring can be healthy, is that cruel? Or is it cruel to let both baby birds die because they can't be sufficiently fed? And if attempting to feed both baby birds caused them to be weak and sickly and perhaps weaken the species as strong genes could not be passed on, is this not cruel? Or would it be cruel to kill all the birds, mother included, because some other species became overpopulated and dominated all life support? Look around the world at the sad condition of about three billion people suffering from hunger and malnutrition—the largest number in history, almost half the world's population—even some in the U.S. Look at the sickness and death that millions of these people and their children endure year after year, and then look at all the other species of plants and animals that we force into an early death and even extinction, both situations because of human excesses. Is this not cruel? And unusual?

Could humans drop one of their children from the nest? Most of us certainly could not and should not, but infanticide (killing babies) has been practiced in the past as a method of controlling population, for instance in Japan and New Guinea. It is even rumored that in recent modern times infanticide—specifically killing female babies—was part of China's population-control policy. For years China also pushed a one child per couple policy and has forced abortions and sterilizations on women. Incentives in the form of money, jobs, housing, food and medicine have also been used to control population. In Europe during the Little Ice Age (1400 – 1850), periods of starvation drove some people to abandon children so they would be able to feed the rest of the family.

Overpopulation doesn't cause all social problems, but it is a major contributor. It is no news that poverty and starvation, both of which sometimes are caused by overpopulation, lead to crime. Human life does get much cheaper when there is too much of it. In many of the larger cities in the developing world, child prostitution, forced labor and children sold for adoption takes place. Children are abandoned by parents who can't make a living and can't cope anymore. Conservative estimates are that worldwide 25 million of these abandoned children exist in cities as "street children," and the number is growing. In some South American cities people have been attacked and robbed by gangs of young children, some as young as 10 years old. In Africa millions of people live on dollars a week, as there are no jobs for them because the resources and infrastructure don't exist to employ so many people. Consequently, wages drop and people get severely taken advantage of and abused. Overpopulation also causes wildlife poaching, disease and illiteracy, and the burning of forests for crops and pasture.

China's harsh population policies came about in part because their population levels turned out to be much higher than they had thought. This is no different from the bird dropping the chick from the nest, although of course by human standards, randomly taking life is barbaric. What might also be barbaric is that human beings attempt to violate the laws of physics and nature by letting their populations get to the point at which they can't be supported, and therefore feel compelled to bring back the balance by doing it nature's way. The definition of "cruel" might also be that we set ourselves up for the possibility of doing things that are less than civilized, especially when we have the intelligence to do otherwise.

We do not have to have problems that conflict with our morals. However, if we keep on overpopulating and overconsuming, in the future in many countries, including the United States, we may see events that will be quite scary as humans frantically try to control our population because of the terrifying realization setting in that nature is about to curb our population for us, or has already started the process. Throughout history people have done dramatic things when food and space shortages have taken place, including murder and cannibalism. Think about your family members starving to death and knowing there is food just over the hill—but only enough for two or three families. But the problem is that 100 families are headed over the hill for the same food. How far would you go to ensure your children did not starve to death in front of your eyes?

In times of extreme trouble we in the United Sates also will do dramatic things. It is folly to test any culture's mettle by letting the extreme happen. And as I said, the extreme can happen, such as very serious food shortages in developed nations (developing nations already experience starvation) if we toy with climate mechanisms enough. When it happens, there is nothing that we will be able to do because most of us don't have the ability to get our own food or other resources. We depend totally on an infrastructure that might not be functioning well when upheaval strikes. If our minds do not accept this potential reality, this is only because, as I have said, we have been programmed to believe that humans are invincible. When the earth and its workings are understood, it becomes evident that everything we depend on really can unravel.

Populations may have exploded above what operating economies can keep up with. It's a giant catch-22. Elevating people from poverty to a higher economic status lowers fertility rates and so reduces some of the negative effects associated with excess population. And educating any population, especially women, has also been shown to reduce fertility rates. However, consumption levels increase dramatically, as people want things and can afford them, so that the effect a lower population has on the planet can be greater than that of the previous higher population that didn't consume nearly as much. That is why from the standpoint of some environmental issues, such as global warming and resource use, the United States, with about 300 million people and a very high consumption-density, is more "overpopulated" than Africa, with close to a billion people. How many people can we remove from poverty—and what level of consumption will they strive for?

If we are going to use any idea that pertains to the economy and education to get the world out of poverty, the economic part of the plan can no longer allow excessive growth. The economic part of the plan must be implemented very carefully, with an eye on careful consumption, conservation, new technology and the environment. Therefore the American way of life cannot be copied, but only loosely mimicked. Even after people are taken out of poverty, their culture must still strive to reduce its numbers as we too in the United States also must do. A new way of looking at consumption and carrying capacity must be established. Business as usual will not work. The earth's ability to absorb the damage that increased consumption-density brings has been surpassed.

Look at China today. The demands it puts on the rest of the planet by importing the resources it needs—wood and oil for example—contribute dramatically to world deforestation and pollution. The problems in China with air and water pollution, soil erosion, waste and resource use are enormous. Unless there is serious change, these problems may become impossible to solve, as China keeps growing its population, economy and consumption levels. China's contribution to climate change is increasing, and it will soon surpass the United States in the emission of global-warming gases. It is absurd to contemplate the additional damage that will take place if the Chinese continue to expand and grow their economy and consumption levels, and base this mostly on fossil fuel use. When we look at India and many other developing nations also on the climb to first world status, the growth is even more irrational and alarming.

As population swells and environmental problems affect our ability to live, human life will get cheaper and cheaper, and if too many people fall into extreme poverty, the rest of the world will drop them out of the nest, maybe not directly but certainly indirectly by turning their backs. This is going on right now because we are far from taking proper care of all of our own species. The food lacking to feed the bird is no different than the resources not being there to maintain the person. It would be so much wiser to control population growth humanely, to develop an economy that works more realistically within the bounds of nature's limits, and then to take care of everyone, as well as caring for all the other species of plants and animals that support us, so that we can be the intelligent, humane species that we are supposed to be.

Imagine it's the year 1830, and there are about one billion people on earth. It took 4.6 billion years to get the first billion here, or certainly at least 600 million years of plant and animal evolution and finalizing ocean and atmosphere evolution. Then, in only 100 years, by around 1930, another billion people show up. Medicine and hospitals were getting smarter. Penicillin and other wonder drugs were taking off. Sophisticated agriculture and refrigeration were starting, and at the same time we forgot that we were part of nature, as we started to think that we could ignore the natural world.

By about 1960 we put another billion people on the planet. What previously took 100 years, and before that at least 600 million years, now took only 30 years. By 1975 or so another billion people—in only about 15 years! The next billion people showed up in roughly 12 years, and now it is 1987 and there are five billion of us. Scientists like Paul Ehrlich yelled at us about all the consequences of overpopulation, but almost no one was listening. Throughout the population increase our politicians and business leaders were telling us that all is fine. Miracles in agriculture and technology will go on forever, and we will be able to support a continually growing population. Meantime, the problems of poverty, overpopulation, consumption and environmental degradation have gotten worse. By 2000, 13 years later the population reached six billion, and in 2008 it hit 6.7 billion. We are heading for seven billion in the next four or five years, eight billion 12 to 14 years after that. Without social change our numbers will exceed nine billion people during the following 25 years, by about 2050.

Fertility rates—the average number of children a woman has over her lifetime—have declined. However, currently we are adding more than the populations of Los Angeles, New York, Chicago, Detroit, Philadelphia, Boston and Dallas combined to the planet every year. That's about 75 to 80 million people a year. Taking birth and death rates into account, this means that the average increase is roughly 2.5 babies born every second! Every single aspect of the planet is under assault by human beings. Damage to systems is everywhere we look, whether the damage is biological or physical.

In 1955, with about three billion people on earth, half the world was living in some kind of poverty. Of today's 6.7 billion, about half are living with malnutrition, disease and/or some level of poverty; there are over 800 million people who are seriously undernourished, and about one-quarter of them are children at risk of starvation and death. It is said that there is enough food produced in the world for everyone, but political conflicts, wars, natural disasters, corruption and poverty prevent the food from getting to everyone. This may be true, but if environmental damage increases and resources become scarcer and more expensive, the number of people in poverty will increase along with all the problems that will continue to keep people from obtaining food.

About 80 percent of the world's wealth (represented as gross national income) is controlled by the high-income developed nations that contain only about 15 percent of the world's population. The lowest-income developing countries control a very small percentage of the world's wealth—less than 5 percent—while accounting for about 37 percent of the world's population. Over half the world's people earn less than $9,000 a year. Almost 2.3 billion people earn less than $800 a year. About two billion people live on subsistence agriculture involving hand labor that marginally meets the food needs of a farmer and his family. One and a quarter billion people live fairly well, in the developed world of the United States, Canada, Europe, Australia, Japan and others. In the rest of the world, developing areas such as China, India, Africa, South America, Central America, and the Middle East, approximately five billion people for the most part don't come close to the developed world's living standards; in fact, two billion of them live without electricity. Over 90 percent of the world's population growth is happening in developing countries.

All this while the U.S., with about 303 million people, has possibly the highest proportion of obesity and overweight people in the world, maladies caused (leaving out medical problems) simply by overeating and lack of exercise, a culturally induced norm. This has created millions of sick people suffering from a range of problems from digestive diseases to diabetes, heart problems, and for many, an early death. Even children as young as 10 have the beginnings of clogged arteries. By some estimates, 118 billion dollars a year are spent in the U.S. on weight-related problems. It is estimated that worldwide about one billion people are overfed. We overconsume many resources and all of this consumption adds to environmental degradation.

Greater consumption causes a greater need for matter, energy and space. Each person needs certain things to live in a modern world. A baby is born, and in his or her lifetime a certain amount of resources such as oil, coal, food, water, wood, etc., and all kinds of manufactured items will be needed; therefore overall environmental damage will be required for the person's survival. What country and lifestyle the newcomer lives in will determine how much damage is done, and what type of damage. Live as a native in a South American rain forest, and almost no damage will be done. Live in a big city in Europe or the United Sates, and severe damage will be done. This damage varies as some areas of the world, such as Scandinavian cities, have pushed towards cleaner, less-wasteful urban infrastructures. The U.S. is one of the more wasteful cultures. With about 4.5% of the world population, the U.S. produces up to half of the world's industrial waste and consumes about a quarter of many of the world's commercial commodities. In one day the average American uses about 40 pounds of fossil fuel, 30 pounds of other minerals, 22 pounds of paper and wood, 27 pounds of farm products and over 75 gallons of water. Collectively, every year, Americans throw away two billion disposable razors, 25 billion Styrofoam cups, 67 billion cans and bottles, 50 million tons of paper and 18 billion disposable diapers. This is a partial list.

Cities contain almost nothing in the way of natural resources, although there might be some water underground. Cover is artificially produced, as it is in suburban homes, but in a more damaging and concentrated way, as the carrying capacity of the area, Manhattan for example, is passed many times over. The population density of a city is very large because more people have been placed in an area than that area can sustain. With over eight million people at any given time, New York's Manhattan Island and surrounding boroughs cannot possibly sustain their population without all the resources provided by the countryside. Land for food growth, land for resources such as ores, wood, oil, coal and water that all need to be brought in and processed into the things that people need to survive. Large cities degrade tens of millions of acres somewhere else in the country and the world. The average American consumption level requires about 23 acres of land per person, while the average for the world is 5.6 acres.

A large city produces millions of tons of waste, garbage that has to be trucked out to a landfill or incinerated, and hundreds of millions of gallons of sewage that has to be treated and disposed of every day. The concentration of oil burners in thousands of apartments and industrial buildings, the cars, trucks, trains and the electricity use and manufacturing all create vast amounts of concentrated pollutants in the air, water and soils. All these pollutants contribute to everything from global warming to acid rain and cancer. It's okay to have a few big cities here and there, but when the whole country—and the whole world—increases the number of cities, and is enlarging every city to the point where nearby suburbia gets eaten up by the expansion, then suburbia degrades. In turn, often the suburban populations migrate further away to eat up rural land and make a new suburbia, which also will eventually be eaten up by the ever-expanding process until too much is urbanized.

The typical suburban neighborhood of densely packed homes, some with lots as small as 40 or 50 feet by 100 feet, many times built row after row, is hardly more capable of taking care of the needs and waste of any home's inhabitants than the city can do for an apartment dweller. Long Island, New York, is a perfect example of an expanding suburbia, with about three million people who have crushed themselves into a city-like suburbia with a decaying quality of life. The world really is round: we've run into each other.

In part because of environmental degradation in poor countries people are driven off their land and into cities to look for jobs that don't exist. As of 2007-2008 urban areas are home to the majority of the world's population for the first time in history. Many cities in the world have shanty towns around them, housing millions of poor people who may never be employed. Look at Mexico City or Rio, to name just two. Look at the United States. Our cities are not as bad as some other world cities, but we have our share of poverty and run down neighborhoods in and around many of our cities. Over 40 million Americans are at or below the poverty level, and 30 million have been classified by the U.S. Department of Agriculture as food insecure. As world population grows and more and more people everywhere are locked into poverty, immigration pressures on countries like the United States will be even more enormous than they are now.

When talking about carrying capacity, we can look at the Caribbean island of Hispaniola, which comprises Haiti and the Dominican Republic. Both countries suffer from overpopulation, which is easy to see because Hispaniola, a small landmass when compared to the U.S., has a clear and indisputable boundary—the ocean. Haiti has serious political and social problems, and a major part of its plight stems from overpopulation and its effect on the ecology and economy. Haiti suffers from massive deforestation, water problems, disease, soil erosion and poverty. Unless there is social change, or nature steps in, by 2050 Haiti's population will more than double to almost 19 million people; then something may be forced to change. There are other islands around the world with obvious population problems.

The United States is an island if we look at it that way. It's much larger than Haiti, but no less subject to destruction than any small island; the destruction simply takes longer to see and realize. The world is an island floating in space and is no less susceptible to any island's problems including complete collapse, if we don't use common sense and anticipate the obvious.

The United States is taking in more immigrants than at any time in its history, and we have the fastest growing population of all the world's industrial nations. Estimates vary from 1.5 to three million immigrants a year entering the U.S. with a very large percentage of them illegal. No one really knows the true number of illegals, but estimates of the numbers already here range from 12 to 20 million, some say more. The U.S. continues to debate immigration reform. However, this government is hardly looking at environmental issues and so far is not asking the American public for their input. Rather, the decision on immigration reform will likely be based on corporate desires and profits, the economy and tax revenues.

Taking age range and the higher fertility rates of some of the newcomers into account, plus our current growth from our existing population, and high-end predictions show that in 20 years the U.S. could have another 80 million people. By 2050, the roughly 303 million people in the U.S. could increase to 450 million, and we could double our population 80 years from 2008. That's about 600 million people. One study put current U.S. population at a net increase of one person every 11 seconds! This country's traffic problems, pollution problems and quality of life stresses due to our current population and lifestyles are already problematic. It is doubtful that the American public would be better off with the added stress of only 100 million more people, and we certainly would not deal well with the stress created by 300 million more people. This would shatter our ecosystems, severely damaging our environment, and by then we would assuredly be out of our own oil. Presumably, if we hadn't economically and socially disintegrated yet, we would have gotten onto alternative energy sources, yet our ability to maintain a quality infrastructure for such a large a population would be very questionable and of course the rest of the world may have also reached some sort of extreme. At this point there would be no guarantee that the world would not already be in serious environmental trouble.

Conservative estimates—lower-end predictions—of U.S. population increase show that if nothing changes, population will increase by about 50 million over the next 20 years and will double over the next 120 years, which is certainly better than doubling in 80 years. So let's go with the longer time frame; it feels better and looks better. So then, the fact is, either way it seems to be an accepted belief that the U.S. can keep increasing its population, maybe even double it. We may never hit 600 million people. Environmental change may stop our population growth, or social change may stop it. However, too often in the system of politics and business, continuing to increase population is still seen as a good thing, and is addressed for the most part under the context of how much more revenue and business increasing population will generate. Since 1900 U.S. population has about quadrupled. Should we quadruple it again to over 1.2 billion people?

There is another problem with U.S. population growth. Remember that worldwide we are responsible for 25 percent of the oil burned (as different products), about 20 percent of the coal burned, and also large quantities of natural gas, as well as vast quantities of other resources used. Environmental damage from the demands we put on the planet is substantial and out of balance with the rest of humanity. Again, 303 million people out of 6.7 billion are contributing about 25 percent of the greenhouse gases that threaten humanity. As we increase our population day by day and year by year, we bring more people into this incredibly excessive energy- and resource-use lifestyle, and therefore we continually and dramatically accelerate global warming, along with other environmental problems. As the world's population increases, only a handful of countries in Europe and elsewhere have actually achieved declining or stable populations.

Immigration is a touchy issue because we argue along the lines of politics and cultural divides, things the earth couldn't care less about. The United States' population growth is out of control, and arguments based on cultural and political beliefs that support immigration only waste time. I've even heard people say it's not fair that their parents got to come here if someone else's can't. Then there are the arguments that we get cheap labor and this boosts the economy. The Democrats wanted votes; the Republicans wanted the cheap labor for industry: now both parties want the same things. So far the discussions about immigration are meaningless unless we look at this issue from nature's point of view.

The earth long ago established how much water, oil and coal are under foot, how much soil, land, trees and all other resources are available for all the various species, how much pollution and damage the earth could handle. The earth didn't make any provisions to be discriminatory and only give resources and stable climate to certain cultures, political groups, or religious groups. The earth simply said, here is what I have: use it up, or seriously change me, and you are all finished. The earth went on further to say that patience was its virtue, that earth didn't really care if humans wiped themselves out from overpopulation and over consumption, because the earth has other species and might make some more.

To say that the earth is talking is as silly as to say that immigration should be based on anything other than the carrying capacity of the United States—or any other society's island of real estate. This carrying capacity must be based on a sane amount of consumption, on conservation and lifestyle change and on new technology—and not on the insane and wasteful way in which we live today. In a sense the earth is talking because clearly it does give us a fixed amount of living essentials and has shown us that we are seriously damaging these essentials; science has identified the processes very concisely. To debate about an immigration policy is utterly ridiculous, without first establishing what the carrying capacity of the area is, thereby first and foremost having a population policy. How many people do you load on a life boat? Well, first you need to know how many the boat can hold, and then you must be very careful not to fill it to the maximum so that you don't tip over! I am far from the first one to try to make this point.

So what is the carrying capacity of the United States? The question is as difficult to answer as the question about the world's carrying capacity for humans. We will never know the exact number. But it is obvious that there is a limit to how many people we can support. It is obvious that stress from traffic and overcrowding is getting worse. It is obvious that we use vast quantities of resources, many of which will run out in the foreseeable future. It is obvious that normal climate change and/or global-warming-induced climate change creates the real possibility of food and water problems. We see global warming, species extinction, pollinator decline, air and water pollution, acid rain, sickness in people, forest and crop damage and massive habitat destruction. We see these things in the U.S. and all over the world, and they are caused by too many people consuming too much stuff. Even if we conserved and went totally over to solar power and eliminated the evils of fossil fuels, many problems would still continue. It is clear that there is a limit to the number of people we can support. It is common sense.

Three hundred million people in the U.S. is too much—and too risky based on what is already happening. We should not want to be at carrying capacity anyway, because when problems arise, we need a safety net. If the U.S. had 200 million people and food problems arose, we probably could deal with the situation. But if we end up with 450 million to 600 million people, then we may not be able to deal with the situation. We need to use science and common sense and call out a smart number, especially considering that carrying capacity fluctuates up and down as the environment changes naturally for better or worse—and we must consider that global warming will make it worse.

Carrying capacity based on a species that functions within the strict Darwinian boundaries that nature sets would mean a very primitive existence for humans. However, we do not have to live without the conveniences of a modern and technological world, but we must stop asking science and technology to attempt to violate the laws of physics and nature—which can't be done anyway—because we are physics and nature; we are intertwined in the definitions of the words.

I will take a chance and walk out on the thin ice again. I think the U.S. would be a much healthier place with a safer future if population was kept at 200 to 225 million people. The lower number being better.

Two and three hundred years ago people came to the United States and ravaged the land, cutting down so many trees in the East that trees had to be brought in from the West. From 1620 to about 1920 the majority of the forest in North America was cut down and replaced by agriculture, pasture and other uses. Animals were slaughtered en masse, and the land and resources were treated as though they were endless—but population was relatively low and there was room for expansion. Even so, a lot of damage was done. By 1920 we knew little about the environment, and more and more people kept coming. But it's not 1920 anymore! Science has studied this planet and its environment very carefully, and the planet's systems are now very well understood. It doesn't matter to the ground under our feet, or to the air and water around us, where the additional people come from. Whether the carrying capacity is exceeded by natural births of all one culture and nationality or exceeded by 20 different cultures and nationalities, the environment degrades. Psychological stress rises and quality of life goes down from overcrowding, and at some point life starts to end.

The people coming to the U.S. often come from cultures with very high fertility rates, some as high as five. Replacement level is hypothetically two (adjusted slightly higher for mortality; in developed countries 2.1; in developing countries higher), which is the number of children needed per couple to keep a population at zero growth. When the two parents die, they have been replaced by their two children.

The United States is generally an educated country with poverty that is certainly not as extreme as developing nations, and consequently without immigration our population growth rate would be considered low—but should still be lowered to zero. Immigration is drowning our country with people given their numbers alone, and given their higher fertility rates these immigrants increase our otherwise almost manageable growth rates, even if the newcomers eventually join the crowd and have smaller families. (The U.S. is not the only country with immigration problems).

So who is monitoring our land? I don't only mean large tracts of federal or state land. Every day trees and other plant life are being steadily destroyed, like squares on a checkerboard, to make way for our increasing needs. We know that all over the country rural areas are becoming suburban as farmers and other land owners sell off their land. From orange groves in Florida to corn fields, dairy farms and forests in North Carolina, and everywhere from east to west, the bulldozers and developers are rolling to make room for a rapidly growing population. It's an individual's choice to sell his or her land, and if I were ready to retire and had land that was suddenly worth a small fortune, I might sell out too. But the concept in question might be viewed very differently, if all who sell their land to developers thought collectively. Then the view might be of a country that is being ruined from habitat destruction, species extinction, pollution and the like, as well as the aesthetic value of beauty and tranquility that is being lost forever, which adds to the psychological stress on people. When there is nowhere left to run and hide, and relax, along comes the chronic stress previously talked about. Chronic stress is something that we all deal with to one degree or another, and certainly where we live in the U.S. may determine how much stress we are subjected to. However, more and more, stress is everywhere, and to at least some degree that stress is caused by overpopulation and by a society that is running too fast and trying to produce and consume too much. As long as the U.S. population keeps growing, there will be people to buy the land and someone to develop it. There will be more homes, roads, shopping malls, golf courses and commercial enterprises. Not only are we damaging the planet, but we are losing all our beautiful little towns and our culture. If we keep it up, there will soon be little peace left for the United States and the rest of the world.

We know that when a country has a higher rate of education and a lower rate of poverty, its growth rates (the rate of natural population increase) are lower. Less than 1 percent is considered low, yet it should be zero. When a country has the opposite, low education and high poverty—especially extreme and widespread poverty—then the population growth rates are high, over 1 percent and as high as 3 percent. Small numbers represent high growth rates because it's like compounding money in the bank. Three percent of a billion people is 30 million, so after one year there are 1,030,000,000 people. After two years, 1,060,900,000; after three years 1,092,727,000, and so on, and after 10 years, approximately 1.35 billion people. A little over 10 more years and we hit the two billion people mark. (Of course this assumes a constant growth rate, which is not always the case. Rates do change, up or down.) This is why the current world population of about 6.7 billion will continue to grow so much with only a 1.25 percent growth rate. Along with fertility rates, growth rates have declined from years ago, but the human J-curve will continue before changing. Even if fertility rates continue down as they are, world population may only start leveling off at about 9 billion people and possibly peak at about 10 billion. If 2007 fertility rates do not change, then population will continue higher, possibly hitting 11.5 to 12 billion by 2050. In a best case scenario, if fertility rates drop further in the near future then population could level off at about 8 billion somewhere between 2040 and 2050, and possibly decline from there. And, of course, a population crash is possible at anytime.

China has over 1.3 billion people; India over 1.1 billion, and among the various nations in Africa, over 900 million. By 2050 China will have close to 1.5 billion people and India will have about 1.6 billion, the African population more than doubling, hitting close to the two billion mark. China and India still have serious education and poverty problems, yet both countries are educating some of their people and are raising the status of women, and many people have risen above poverty. This is lowering growth rates, and although it may be too little too late, yet certainly it is better than no action. On the other hand, surprising social change could push growth rates lower.

As new consuming nations, China and India are using vast amounts of energy, even though only a small percentage of their population have become more modern consumers, and they are trying to increase that number of consumers dramatically and mimic the U.S. economy. Even if China, with about 20 percent of the world's population, stopped growing at 1.3 billion, it would still have over four times the current U.S. population. Some business leaders love this growth scenario, because as China's and India's new, potentially gigantic consuming class consumes more and more, developed countries like the U.S. get inexpensive manufactured goods and services that are imported and outsourced. Worldwide production and consumption increases, lots of money is made and the rich get richer. In 2000 approximately 2.8 million people, about 1 percent of the U.S. population, had as much to spend after taxes as the bottom 100 million Americans.

The United States must stop playing politics with family planning and birth-control advocacy organizations such as the United Nations Population Fund. This politicking has gone on for years. The UNPF works in approximately 146 countries and territories to help spread contraception methods and reproductive health care for women. The Bush administration has withheld over 120 million dollars of funding, the only country to do so for reasons other than budget. Those reasons are the politics of religion, abortion and birth control. The result of years of withholding money has been millions of additional abortions and unwanted pregnancies, thousands of maternal deaths and untold suffering of women and children in developing countries.

The world is very aware of overpopulation problems. The United Nations and the World Health Organization are not strangers to the issue, and efforts are being made to stabilize population. However, much more urgency needs to be placed on the issue, and the United States needs to act like the progressive, educated country that it should be. The United States cannot completely control world population issues, but it should set an example for the world by at least stopping its own population growth. All immigration must stop—with exceptions of course—while family planning and birth control must gain in importance. The U.S. should have a two children policy—zero population growth—that comes about not through telling people what to do or by enforcing anything. Instead this new cultural norm should come by way of education and common sense. And the population should be allowed to slowly fall as some people do not have children or prefer to have only one. Without immigration U.S. population might very well start to decline on its own, as our fertility rates are low, and may head lower.

The U.S. should export birth control technology to poor countries that need it. All developed countries must take similar action (much is being done), and all countries must manage their own populations. Certainly, help to countries with overpopulation pressures should always be offered, but if a country overpopulates, then the population belongs to that country alone. Overpopulation in any country is simply too dangerous to ignore.

Either we control our population growth sensibly and humanely, or we wait until nature does it for us—because nature will, one way or another, bring down our numbers.

7
DISEASE

Disease is as natural as the wind and can come from many sources, and bacteria or viruses will go after any animal, including the human one. Worldwide, millions of people die every year from AIDS, diarrhea, respiratory infections, tuberculosis, malaria, tetanus, hepatitis, intestinal problems and a host of other illnesses. Infectious disease alone claims the lives of thousands of men, women and children every day.

The developed world benefits the most from modern medicine and technology, while people in many developing areas such as Africa and Asia, living in more of a Darwinian realm, still succumb to the struggle for survival. Disease, starvation and poverty keep many people living more by nature's rules, even with help from charities. Developed nations like the United States have their share of disease, but the disease is minimized because of money and quality infrastructure, and some diseases have basically been eradicated—but all this can rapidly change.

As population increases, disease comes more and more into play. It is as though nature is trying to stop human numbers from increasing. Poverty and catastrophic events also add to the problem of disease. With events such as a volcanic eruption or a hurricane like Katrina, when large numbers of people die and many are left without any sanitary services and polluted water abounds, cholera and other bacterial events can arise. However, we don't need catastrophes for disease to lower our numbers. Overpopulation breeds poverty, and poverty breeds overpopulation—and poverty breeds disease, as masses of people become congested in large and small cities with unsanitary conditions and poor health from lack of proper diet and health care. It has even been shown that poor people suffer from the psychological stress of despair and worry from living in poverty, which then lowers the immune system's ability to fight sickness and ill health. Poverty and overpopulation put more people in contact with animals that can transfer disease, as with the avian flu in places like Vietnam where people are in unsanitary contact with chickens and other birds. The closer people are in contact with each other, the easier it is for disease to spread once it gets started, even in developed countries like the United States where there are so many densely populated cities.

Infection may be caused simply by humans stumbling onto disease as our numbers increase and we spread out into areas previously too remote for most people to get to. This is how AIDS may have come out of Africa. Over the past 25 years, as population rapidly climbed, approximately 30 new and re-emerging diseases, many deadly, have arisen. These diseases have chased us at the unprecedented rate of about one per year for the past 25 years. A short list includes: SARS; Ebola; AIDS; Marburg; West Nile; Mad cow; H5N1-Avian Flu; Lassa Fever; Hanta Virus; drug resistant Malaria and Tuberculosis; Monkey Pox; Lyme; Staphylococcus aureus, and various forms of hepatitis. As global warming continues, disease problems will get much worse, in part because warmer temperatures allow disease-carrying insects to thrive and expand their range. Today malaria is affecting a record number of people and may threaten half the world's population.

Mad cow disease spread in England's cattle herds in the 1990s because cows were fed parts of cattle carcasses mixed in with feed. At least 100 people died from the disease. The human equivalent of mad cow, called Creutzfeldt-Jakob disease, has been seen in Papua New Guinea, where cannibals eat the brains of other humans. A frightening aspect of these brain-wasting diseases is that it can be decades after infection before the disease is detected. There may be many more cases yet to arise. The United Kingdom has banned the practice of cannibalistic feeding in order to halt the spread of mad cow disease to people. The European Union has banned giving cattle, chicken or pigs any animal protein in their food. In 1997 the U.S. did something similar, but made exceptions to the rule. Cattle blood products and fat were still allowed to be fed to cattle. Pig and fish protein, feather meal and even chicken manure were still fed to cattle, while the chickens and pigs were fed cattle protein. Some worry that this is a back door for mad cow to still get to people. The process was supposed to be stopped, but it still continues.

Some diseases may not actually be new, but only new to us, as might be the case with HIV which causes AIDS and has killed about 40 million people, and may kill another 60 million over the next 20 years, depending on treatments and if these treatments make it to poorer nations. Currently there are about 33 million people infected with the virus.

Diseases such as malaria can be linked to something as subtle as deforestation, because cutting down trees changes the local environment and adds to the number of puddles in which mosquitoes can breed. In most developing nations, when people migrate to cities en masse, they bring disease with them as well as helping disease carrying insects to thrive in crowded shanty towns. This can happen by something as simple as creating gardens or stagnant water situations in which insects will breed. Large quantities of human waste can become concentrated and unsanitary, also spreading disease. Millions of people worldwide have no sanitation or clean water, and about two million people die every year from diarrhea caused by drinking contaminated water, most of these deaths occurring in children.

Poverty and poor living conditions cause disease to spread, but we in the United States are not completely safe just because we live in a developed, sanitary country. Once some diseases get going, there are no boundaries, especially if the human world creates the ideal medium for the disease—and some diseases can easily spread in perfectly sanitary conditions.

It is possible that AIDS and Ebola may have been dormant for a very long time, and there may even have been outbreaks of these organisms many years ago that we didn't know about because the sickness was buried deep in a jungle of Africa. Yet these diseases did not overtly arise until populations in Africa got larger, roads were cut through previously untouched jungle, and commerce increased. The genie was let out of the bottle, so to speak.

AIDS (Acquired Immune Deficiency Syndrome) was originally called GRID (Gay-Related Immunodeficiency Disease) before the term AIDS came about. When GRID made the move to the U.S. heterosexual population in large part through bisexuality and drug use, the name was changed to AIDS. It was during the 1970s when the disease showed up in the United States and was initially seen in gay men. Some doctors were in a panic, thinking they were witnessing the beginning of another bubonic plague (Black Death or Black Plague, Figure #1) that killed between one-third and one-half of Europe's population during the 14th century and also devastated Asia. However, simultaneously in Africa, heterosexuals were getting AIDS for different reasons that still involved blood contact. Some of the reasons included lack of sterilization of needles in hospitals, high levels of prostitution where very few use or can afford condoms, injecting people with vitamins or antibiotics for profit reusing needles, funeral ceremonies that involved touching the body, and eating animals that carry the disease. Today, AIDS is devastating Africa for many of the same reasons including people not being able to afford drugs to stop the transmission of AIDS from mother to unborn child.

Some viruses rapidly mutate. When we gain resistance to one form of the flu (influenza), nature whips up another one to get us. As of 2008 H5N1, the avian flu, or "bird flu," might still be a threat. This flu is very often fatal. So far the transmission from animals like chickens and ducks to people is rare, but some experts have been warning that it may only be a matter of time before this flu mutates into a strain that will transmit easily from people to people. The regular old flu that we are all used to infects 30 to 60 million Americans each year, and something like 36,000 die. If the bird flu—or another new disease—goes this route, we are all in trouble, as millions could die in the U.S. and several hundred million or more could die worldwide. Even with a vaccine—an Australian company has developed one for the H5N1 Indonesian strain—millions could die, as the H5N1 flu could run in waves from season to season and could last for years.

The United States is not fully prepared for an epidemic of such magnitude. Be it H5N1 or some other disease, eventually we will likely experience a serious disease problem. Hospitals and care givers will be overwhelmed, and millions will be left to die. Economies will collapse as travel, trucking, shipping, airlines, imports, shopping, schooling and much manufacturing slows or grinds to a halt. People will lock themselves in their homes as paranoia runs through societies and a general loss in the faith of government and authority takes place. People will refuse to go to work as the fear of human contact escalates, and many people won't be able to pay their bills. Globalization of economies has broken down biological barriers for disease and makes it even more likely that this type of event will take place sooner or later.

Consider the great flu outbreak of 1918. Often called the Spanish flu, it killed rapidly by affecting the heart, brain, liver, kidneys and lungs. Societies were isolated by the slow speed of oceangoing vessels because there was no jet travel or significant economic globalization. Yet over several years the pandemic killed about 50 million people worldwide—no one knows the true numbers. In our modern, fast and small world, H5N1, something like it or something much worse would spread like gasoline on a wildfire. The 1918 pandemic will look mellow by comparison.

Because of the nature of the virus, it is unlikely that HIV will become an airborne contagion (but never underestimate a virus). However, an airborne superstrain of some other new, highly deadly and very contagious virus would mean the end of humankind as we know it. Three or four billion people could develop the new, fatal disease. This would cause a permanent economic depression and change society beyond recognition.

Doom and gloom some say? Look at Ebola, a virus that makes HIV look like a walk in the park, named after the Ebola River in central Africa. Ebola is a hemorrhagic virus, causing the victim to bleed from every orifice of the body. In two or three days after infection you are on your back as all the body's organs are attacked and some start to liquefy. The painful and tortuous death takes as little as 10 days, and mortality rates can be over 90 percent, as almost no one survives it. Ebola is a hearty virus—more so than HIV—and can live outside the body. Ebola is very deadly, much more so than the flu, and just touching an infected person can spread the disease—although it has not yet spread like the flu. This is because it is not yet airborne, and also because its victims die so quickly they don't have time to walk amongst us and spread the disease as they do with the flu. As the virus mutates, this can change. During outbreaks the military has surrounded villages in Africa with orders to stop anyone who tried to leave the infected area.

So how about an airborne strain of Ebola, the most deadly virus known to humankind? We already have one. It came to light in Reston, Virginia, in 1989 at an army research facility. Monkeys that came from the Philippines were being used for medical research. It is common practice in the U.S. to import monkeys for this purpose but monkeys can carry both HIV and Ebola. There are simian (monkey or ape-like) strains and human strains of both diseases. This time we were very lucky, as the airborne Ebola that the monkeys brought into the U.S. was a simian strain that only killed monkeys, but it was airborne, and it did infect monkeys throughout the building via air ducts. The military and scientists representing a SWAT team of bio-hazard types were called in to kill all 400 monkeys and disinfect everything.

So far Ebola has remained a localized problem with few deaths as compared to the flu or HIV. But will Ebola surface again as a contagious airborne strain—but this time one that kills humans? It's entirely possible. I'll give you another example: the virus that came into Marburg, Germany, some years ago and surfaced in a research laboratory, where it quickly killed several workers. The Marburg virus turned out to be a cousin of Ebola. What new diseases will nature whip up in the very near future? I don't know, but there will be more; just look at the pattern.

SARS (Severe Acute Respiratory Syndrome) is another virus recently thrown at us, and it showed us the ominous problems of our modern way of living. The disease first surfaced in south China in November 2002, and by February 2003 it was spreading across Asia and into Canada. Then in a matter of months SARS spread to almost 30 countries around the world, in large part because of jet travel and the movement of people. Compare this to the 1918 flu pandemic that took years to spread. SARS didn't turn out to be the bubonic plague of the twenty-first century, but it was a wake up call as to how fast disease can spread, especially with our help. SARS could have been a nasty variant of Ebola.

Today even bacteria are starting to run wild and new antibiotics have to be constantly made to keep up with bacteria's insatiable desire to become resistant to them. Bacteria that are resistant to antibiotics are a major problem. There are MDRs or multiple-drug-resistant bacteria, that are resistant to many different types of antibiotics, and there is even an MDR tuberculosis.

Sometimes patients are treated with multiple antibiotics or with experimental ones that may or may not work. People go to hospitals to get well and end up dying because of bacteria that are antibiotic resistant. Hospitals are a breeding ground for bacteria, for example, antibiotic resistant Staphylococcus aureus, which is becoming a major problem in the U.S. In October 2007, the American Medical Association estimated that this infection hit about 94,000 Americans in 2005 and that more than 18,000 died. Even people who go to hospitals to visit sick friends and relatives can become severely ill. The problem is far greater than many in the medical establishment want us to know. People have died from bacterial infections that a doctor could not diagnose in time or simply because an effective antibiotic for treatment could not be found. But the death report might state the cause of death to be complications due to surgery. Not to zero in on hospitals, however; bacterial infections can be picked up anywhere.

A fast way to allow an epidemic or even a pandemic (a pandemic is an epidemic over a larger area) to happen is to overpopulate a country, allow its cities to grow and expand and become very crowded, allow poverty to grow and allow unchecked legal and illegal immigration, especially from countries that have problems with disease. Throw in economic globalization and unlimited travel to and from developed and undeveloped areas, and the mix is complete. All of this is a great way to rapidly spread a new disease or rekindle an old one, especially if the newcomers work in restaurants and in food preparation. Welcome to the Unites States, a developed country doing all these things.

Since 2001 in the U.S. there has been a dramatic increase in biodefense research. The number of biosafety level 4 labs that study deadly and incurable pathogens has also increased, and accidents as well as violations involving infection have taken place. A hearing was held in October 2007, by the House Energy and Commerce Committee to address the problems. There is concern that something dangerous may get out of a lab. The critters that we need a microscope to see can be the most formidable opponents, ones that can be impossible to beat. Disease could wipe out a significant portion of world population more efficiently than a third world war.

Over the years I have polled almost all my classes. "How many of you know someone with cancer, or someone who has died of cancer?" I ask. On average about 65 percent of the hands go up. The American Cancer Society says that one in three females and one in two males will get some form of cancer in their lifetime. How much sickness comes from our pollution and destruction to the environment is not fully known, but it is substantial. What we eat, breathe and drink is having an effect on us. Chemical residue—pesticides, etc.—is found not only in water and on food, but in human breast milk. In 1970 brominated tris and chlorinated tris were removed as fire retardants from children's clothing—from which they were absorbed into the skin and body—because they were found to both be mutagens (causes mutation) and carcinogens. Yet today chlorinated tris is the second most used fire retardant in furniture. Other chemical fire retardants are used in everything from mattresses to clothing, and many of these chemicals bioaccumulate in the human body. Additionally, it seems that fire retardants may not give the fire protection claimed.

Chemicals such as dangerous pesticides that are banned in the U.S. have been manufactured in the U.S., then shipped to countries in Latin America and Asia, only to come back to the U.S. in the food we import. It is believed that all Americans (other societies, too) carry trace amounts of industrial chemicals in their blood.

Pesticides (herbicides, insecticides, etc.) and many other chemicals are released into our environment on a regular basis. (worse in developing nations, such as China.) Possible endocrine disruption from some of these chemicals, and also from the manufacture and use of some plastics, including the leaching of chemicals from plastic bottles and containers into food and drinks, is causing serious concern. The endocrine system produces and regulates hormones in the body that control sexual maturity, sex drive, brain function, growth and development, as well as other functions. Some chemicals mimic the female hormone estrogen and it is thought that they feminize and chemically sterilize some male animals in the wild such as alligators, fish and various birds. The problem may be widespread. Worldwide in many areas since 1938, sperm counts in men have dropped by varying degrees, and it is suggested that endocrine disruptors are the reason. Rates of testicular cancer, undescended testicles and genital birth defects in men have risen. It has also been suggested that the increase in female breast cancer may be due to endocrine disruptors. Phthalates—used in medications, plastics, cosmetics, perfumes, food packaging, and many other products—have been associated with reproductive problems and birth defects.

The use of pesticides on farms has been studied, and the farmers have been shown to have higher rates of Parkinson's disease, prostate cancer and non-Hodgkin's lymphoma than the rest of the population. Their children have shown higher rates of birth defects. Children are very vulnerable to environmental toxins, much more than adults. Chromated copper arsenate (CCA), used to treat lumber to make it rot-resistant, has been shown to be a hazard for children. Is it a coincidence that childhood rates of asthma, cancers, birth defects and developmental disorders have all increased substantially?

I had a complex blood test done on myself recently and trace levels of mercury, lead, aluminum and arsenic showed up. When human blood and urine are fully examined with much more expensive tests, costing as much as 15,000 dollars, many other things are found. In one study, written about in the October 2006 issue of *National Geographic* magazine, blood was taken from an American journalist, and the results were disturbing. Dozens of chemicals were detected, some at low levels, others at alarmingly high levels. Many varieties of pesticides were found, one of which was Dieldrin, known to cause neurological and kidney problems. PCBs (polychlorinated biphenyls) were found. These chemicals were used in electrical transformers, as sealants, in plastics, as paint additives, and in carbonless copy paper. PBDEs (polybrominated diphenyl ethers), commonly used as a flame-retardant, were also found. BDE-47, another flame-retardant, was found. Dioxin, a deadly poison and carcinogen, was found. Phthalates were also found, as well as mercury.

More research needs to be done on the links between environmental contaminants and illness. There are many links suspected, and some have been proven. Components of dry-cleaning fluids, toilet deodorizers, paint thinners, wood preservatives, cosmetics, plastics, gasoline, pesticides, fungicides, flame retardants, non-stick-surface chemicals, as well as many other industrial and household chemicals, do not belong in the human body, and neither do heavy metals. They do not belong in the system of any living thing.

8
US AND THEM

We hunt them for food and for sport, abuse them, poach them, destroy their homes, change their climate and put them in cages for entertainment. The net result is the beginning of a mass extinction crisis right now. Species all over the world are going extinct, and the threatened and endangered lists are getting longer (threatened means at risk of becoming endangered, and endangered means imminent risk of extinction). As already mentioned, 20,000 species a year may be disappearing. When a news commentator talks about losing another species, the loss is often viewed as an unfortunate result of the growth and expansion of the human race, and so although there is an aesthetic value lost, it is one that must be tolerated. However, there is a critical point when enough of them—plant life (flora) and animals (fauna)—go extinct, after which we will have problems with our own survival. We don't know where this point is when ecosystems and food crops breakdown, or the atmosphere and oceans change surprisingly, for example, or to what degree these things can happen—but we may soon find out.

A keystone species is critical to the survival of many other species and the ecosystem in which they live. For example, kill most wolves—as we practically speaking did do, slaughtering two million more or less since settling North America—and an entire ecosystem is changed from the demise of just one species. Elk—the prey of the missing wolves—overpopulate, then aspen trees, cottonwood trees, willow trees and shrubs decline because of overgrazing elk. Beavers in a particular area may diminish because as the trees decline they lose a food supply and building materials. No beaver dams means fewer succulent plants, and so grizzly bears may decline or leave the area as they lose a critical food supply, especially after hibernation. No beaver dams means fewer ponds and fish, causing another effect on the ecosystem. No elk carcasses left behind after the wolves get their fill means there may be fewer bald eagles, coyotes, magpies and ravens, all of which act as scavengers. If coyote numbers drop, then their prey, rodents such as voles and mice, explode in population, causing environmental damage in other ways, such as an increase in the fox population that feeds on these rodents, and then a decline in birds that the foxes also prey on.

In marine systems similar problems arise. Sea otters are considered a keystone species. In Alaskan and Aleutian waters from the 1970s to about 2002, along a 500 mile coastal area, sea otter populations dropped by almost 90 percent, from about 54,000 to 6,000. Extensive whaling was eliminating so many whales that their prey, zooplankton, increased considerably in numbers. This caused an explosion in the pollock fish population because the pollock fed heavily on zooplankton. However, this took away a food supply for other fish like ocean perch and herring, which then declined along with, eventually, the pollock. This loss of food supply caused harbor seal and Steller sea lion populations to crash. Consequently, killer whales (Orcas) now lost a major food supply, so many switched over to sea otters, which in turn crashed the sea otter populations. Without sea otters as a predator, sea urchin populations dramatically increased, and they in turn ate up so much kelp that this ecosystem diminished. This caused a collapse in other fish species and marine organisms that relied on the kelp forests. It was indeed a chain reaction, and in August 2005, sea otters were put on the threatened species list. Clearly, ecosystem functioning is a finely tuned, very complex instrument, one not easily played.

Obviously it is difficult to say exactly what will happen when an important species disappears, and the events that led to sea otter decline, or changes that took place from wolves declining, might be different than what was just described. Research is ongoing, and debates arise. However, it is a certainty that something will happen when we diminish a species. It may be clear that elk numbers will increase without wolves, and indeed within 10 years of the reintroduction of wolves into Yellowstone in 1995, elk numbers dropped, while aspen, willows, cottonwoods and beaver populations increased. But still, it is not always easy to figure out the complete puzzle.

We also must take into account that ecosystems will always re-establish a new balance when damage is done. The forest may survive without the wolves, especially considering that we have eliminated so much habitat; trying to bring back the original balance might be impossible, and reintroducing wolves, as was done in Yellowstone in 1995, may actually be a new disruption. I am not saying that this is the case, but when we mess with a system, repairing it to a level we want or need can be difficult or impossible. If a few keystone species go extinct, ecosystems will adapt, but if too many keystone species vanish, radical and dangerous changes will take place. The world's ecosystems will still adapt, but many other species will become unsupportable as life-support mechanisms fail.

By no means am I implying that the species that aren't keystones aren't important. They all have their functions, including supporting the keystone species. Any one plant or animal that we look at anywhere in the world, even in our backyards, if driven to extinction or dramatically reduced in numbers would then cause some other plant and or animal to be affected. All plants and animals are connected like a maze of chains, all hooked together in a tapestry or gigantic food web. Break out enough of the chain-links and the tapestry will assuredly tear apart.

Trees and other plants transpire vast amounts of water (more on this later), which puts moisture into the atmosphere to help maintain clouds, rain and climate. Clouds affect how much of the sun's energy reaches the planet's surface. They reflect, absorb, and reradiate energy, all vital to specific climates. Reflection is called albedo, and the planet has an average albedo of about 30 percent, meaning that this much of the sun's energy is reflected back to space, while 70 percent is absorbed. Average albedo is based on the reflectivity of the earth's surfaces, such as forests, dirt, grasses, water, snow and ice, as well as the atmosphere and clouds. Trees themselves have an average albedo. Cut down trees or destroy any plant life and expose something else, be it dirt, grass, concrete or other human structures, and albedo changes.

Forests and plants lessen water runoff and thus minimize soil erosion and maintain nutrients for continued growth. This also helps recharge aquifers so we have drinking water. And again, we get our oxygen from plant life, part of which is floating phytoplankton in the oceans. This phytoplankton is the base of the food chain in the oceans.

Without it most marine life would vanish, along with about half our oxygen supply. Plant life supports animals like the pollinators—the insects, birds and bats that put half our food in our mouths—and the pollinators support the plant life.

Animals spread nutrients with their fecal matter and urine, as well as in their life and death cycles, and they spread uncountable numbers of seeds, both processes causing things to grow. The wild food plants that we depend on, as well as many cultivated crops, are dependant on plants and animals supporting each other. Disease-carrying insect populations will increase if the animals that feed on them diminish, and insects that damage plant life may also increase in numbers.

We get fibers and wood from things that grow, and many developing nations rely on wood for cooking, heating and of course shelter. Many of our drugs contain chemical compounds from the wild. Aspirin was first synthesized from the bark of willow trees. There are undoubtedly cures for diseases out there that we can have, or we can kill them off with the extinction of the species yet to be discovered. There are countless and endless symbiotic relationships between plants and animals on land and in marine environments. Kill off one and we kill off the other.

If we were to maintain an average extinction rate of 20,000 species of animals and plants a year and if the decline remained linear, then in 50 years one million species will be gone. But there is no reason why we couldn't face a dramatic surprise when one million species go extinct in less than 50 years, as the dominoes fall faster and faster due to the interdependency of species. Whatever the true species extinction rate is, the pace of death does not have to be a constant one that is linear, rather it can lead to a rapid, exponential increase instead (see Figure #4).

Global warming is causing climate change, and this will dramatically hasten species extinction and may very well cause runaway, exponential extinctions as plants and animals cannot adapt quickly enough to survive. As we rapidly change the climate, trees and other plants can't rip themselves up by their roots and run to a different latitude in order to get back to a survivable climate. Seed dispersal mechanisms need time to do their job. Animals have trouble with this, too. Some may move, but others won't know where to go, and many others can't cross highways, housing developments, dams, golf courses, shopping malls, etc. Even if they could, the whole balance of predator/prey relationships will be thrown off, as well as the species having to adapt to different soil types and food types. Even breeding can be affected.

Changing the climate is a perfect way to cause a catastrophic mass extinction crisis.

The ultimate symbiotic relationship is our dependence on all of them. The American Indian as predator killed bison, and they grew crops to stay alive, and so they were involved in a food chain. Their food supply was an agreement between the earth and the Indians. They worshiped the sun, wind, rain and wildlife because they understood the importance of these things as manifestations of their gods. Today some would consider this type of worship silly, even stupid, so we pray to our own gods. And if something goes wrong with our life-support mechanisms—the sun, wind, rain and wildlife—will we then pray to our gods to fix the problems that we have created?

The white man slaughtered the bison—three to four million in 10 years and over time many more, no one knows the true numbers—and did so for sport, for their tongues and hides; sometimes the carcasses even were left rotting in the prairie sun. Moreover, bison were killed in quantities that would drive the animals to extinction in order to starve the Indians to death, and the Indian's land was taken and ruined. The Indians could not identify with these acts. Were the American Indians the ideal environmentalist society? Had they been left alone, would they have overpopulated, exhausted their resources and destroyed themselves? Was the harmonistic view of the Indians too fanciful? We will never know, but traditional Indian cultures did embody a basic, sustainable connection to nature.

Most people in developed countries such as the Unites States too often think in terms of their own comfort, needs and desires, many of which far exceed pure survival requirements (not that we shouldn't have some fun). And I am not invoking any biblical prophecy or ideas, but it would seem that many of our faults stem from greed, gluttony and self-centeredness. If a species is going to survive, an individual of a species cannot think or function as an individual, with the only concerns being that individual's family unit and all the safety and comfort that can be gained for that family unit. It would seem that the other animals function as self-centered family units, but knowingly or unknowingly—who really can say—they contribute to all the family units of all species, including the abiotic (the nonliving world of air, water, soils, etc.). Do our superior minds, which give us the capability to have emotions and advanced thought processes, cause us to do the extreme and thoughtless damage to the planet that we do? The bottom line is that we must stop doing the damage and must start to look to them, the other animals (as well as plants), as something to mimic—civilly of course.

Dogs are loyal creatures and sometimes will fight to the death to defend their human companions. There was a case in Japan of a dog continuing to go to greet his master after work for years after the man was dead. A statue of this dog was erected as a symbol of loyalty for humans to aspire to. Are dogs just dumb and blindly loyal to a pack mentality, or do they really feel some form of love and caring? The dog probably looks at its human master as a member of its pack, but I believe there is also some level of love and affection that the animal feels.

Why do so many people love their pets, especially dogs and cats? How is it that animals make us feel good? How is it that a walk in their realm, the forests or mountains, or a swim in the ocean makes us feel so good? The goodness we feel is in part because we are animals ourselves, so watching other animals and making contact with them brings us back to where we came from, maybe back to our roots—where we sometimes belong, where there is simple, truthful balance and understanding. When we look into the eyes of an animal, the connection is pure and real. Yes, the animal is on a different intellectual plane, and on very rare occasion a dangerous one, but the view is uncluttered. There is no greed, hate, deception, etc., like there may very well be with an eye-to-eye experience between humans.

Dogs and other animals can cause complete strangers to bond. Put a crowd of people in an elevator, and there will be tension because everyone's space is invaded. The closeness makes many people nervous. Yet if someone on the elevator is holding a cute little dog, then watch a different story emerge. Suddenly there will be a comment on the beauty and wonder of the pet, and then someone else will jump in. Everyone will smile and relax as safe common ground is found.

When the sick, especially the elderly, come in contact with a dog, patients many times come alive. People have been brought out of silence and depression by hugging a dog. It has been said that contact with animals can lower blood pressure. Reading problems, stuttering and shyness in children have been treated and cured with programs using dogs. In Nevada and other states prison inmates are paired with dogs and, in another rehabilitation program, horses that they care for and train. These inmates show marked improvement in their behavior and attitudes. Even miniature horses have been used to help heal people's minds and spirits. Seeing-eye dogs make and save lives, and dogs are also trained to assist the handicapped, especially people in wheelchairs for whom books, pens and other everyday items must be carried or retrieved. Even monkeys have been trained to feed and care for invalids. The human-animal bond can sometimes be greater than the human-human bond.

People will go out of their way to rescue an animal, be it a dog, horse or cat. Television shows have taken advantage of this reality and put together regular episodes of animal rescues from deep wells, tunnels, holes, out on the ice or in fires (and from abusive humans). People have some sort of instinct when it comes to animals in trouble. They can't stand to see suffering in an animal that may be a pet, or just has some beauty and grace. Yet the same people have no idea that their own actions in everyday life are killing animals around the world.

We teach our children when they are young that animals are thinking, loving, caring creatures. We do this in fairy-tales, cartoons, comics, books and movies. We buy our kids stuffed teddy bears, bunny rabbits and other toy animals of all kinds and teach children to view them as human-like. We even use animals as symbols of strength and safety, like the mighty eagle and Smokey the Bear—the latter prevented forest fires. Ah, but alas, as our kids get older, none of this stuff matters anymore. It was just kid's stuff, not even a lesson about respect for animals or people. Our children, now adults, may go out and hunt animals or destroy habitat directly or indirectly and so destroy the animal's home—and consequently the animal—especially if doing so makes money.

Animals give us what we can't get from human beings without jumping through hoops, without taking great emotional risk and potentially being hurt. Many animals make us open up emotionally, but of course some, like dogs, obviously do this more than others. Clearly an attacking bear will make us run, and sometimes we should run (its generally not a good idea, they can outrun us), or play dead, or stand and fight, or shoot if we have a gun, depending on the circumstance. But even people who are interviewed after a bear or shark attack many times don't hold a grudge or hate for the animal. People instinctively understand what animals are about.

If we compare the scores in the game of death between people and the other animals, we win hands down. Any veterinarian will tell us that animals feel pain. We know this just by observing our pets. Studies have shown that even fish feel pain. So why are there so many of us that are indifferent to animals' pain—or actually get pleasure from it? Sport hunting and fishing are national and worldwide hobbies. Yet both activities involve cruelty and thoughtlessness to animals and sometimes ecosystems. Some people even pay big money to fly to Africa and shoot and kill big game. Many times the animal doesn't die on the first shot so must be put out of its misery with a second shot up close.

Sometimes animals such as deer overpopulate an area, and so it is deemed necessary for them to be culled because human-animal contact can create problems. However, an animal that overpopulates in many cases is doing so because the animal's natural predator, such as the wolf, has been eliminated by us. Further, the problem may not only be overpopulation. Humans destroying habitat—with all our roads, homes, commercial and industrial enterprises—can cause the animal to get pushed into smaller and smaller spaces and into contact with humans.

Some say that it is natural for humans to be aggressive and hunt, even for sport. It has further been said that we are on top of a food chain and so hunting is normal because we are a top predator. Well, top predators such as wolves are involved in a balanced, sustainable food chain in which energy is given and taken. They do not attempt to destroy their food chains, and therefore themselves. Collectively, human beings are destroying not only food chains, but also the entire food web made up of all the food chains. Misguided human practices and attitudes and our aggressive ways are destroying many things and may eventually lead to the last big world war.

Fishing tends to be viewed as a more benign sport. But hook removal does physical damage to the fish, many times ripping out part of the mouth or throat. Some fishermen proudly claim that the fish is thrown back. Does the fish always survive after it is thrown back? And what about all the fuel burned, oil spilled, water and air pollution and contribution to global warming from using a powerboat? (There is certainly room for the sport of boating, but again, the problem here lies with too many people and too many boats—back to overpopulation.)

In the U.S. the processing of animals for human consumption is dominated by large industries. The animals are treated like manufactured objects, in industrial confined animal feeding operations (CAFO) where they are treated barbarically for months, penned up, crippled, branded, and have their teeth, tails and beaks cut not only while the animals are alive and conscious but also without anesthesia. They lead miserable, painful, unnatural lives, often without light or freedom of natural movement. Many animals cannot even groom themselves. Pigs are sometimes kept in crates and are unable to stand or turn around for much of their lives. Hens become deformed and crippled from living in extremely overcrowded wire cages.

In feedlots the cattle are held in overcrowded quarters and fed hay and grain mixtures, much of it corn, which is not a food they are adapted to digest. They want grass, but some say to maintain enough land area for grass might not be profitable for large scale cattle production—plus a cow grows faster on corn mixtures than on grass. However, others say that the humane practice of pasture-raised animals for human consumption is not only environmentally sound for the earth and our health but is also profitable.

The total amount of fat in red meat is much less if the animal is raised on a grass diet, because the corn adds unnaturally to the fat content, a fat content that gets us sick. The grain diet can cause severe health problems for the animals, especially when they are young calves, and the animals are also crowded into unsanitary conditions; therefore they are given daily antibiotics, a practice that significantly contributes to the antibiotic-resistant strains of bacteria that can sicken or kill people. E. coli bacteria in our meat comes in part from the cows wallowing in their own manure, and acid-resistant strains of E. coli caused by the animal's unnatural diet—E. coli that can now survive our stomach acids, whereas before it could not—have also developed. E. coli is deadly. Antibiotics are also used to promote excessive growth along with growth hormones. Over half of the antibiotics used in the United States go to animals.

We ingest the antibiotics and growth hormones given to the animals when we consume animal products. The antibiotics and hormones have also been found in animal waste and in waterways downstream from feedlots, and contaminated fish have been found exhibiting abnormal sex characteristics. In 2007 U.S. legislation has been introduced to partially regulate the use of seven specific feed-additive antibiotics. However, ahead of the U.S., in 1998 Europe banned the use of growth-promoting antibiotics in their meat. Maybe part of the reason for that ban is that the World Health Organization says that indiscriminate use of antibiotics in agriculture poses a significant health threat to humans. The European Union also banned the use of synthetic growth- promoting hormones in meat because of health risks to humans. In the U.S. not only are antibiotics still used, but growth-promoting hormones as well as hormones to increase milk yields in dairy cows are still allowed. In more traditional family farms no drugs are used because the animals are raised in a healthier manner under more natural conditions. Some specialty companies don't use drugs and at least try to let the animals lead more normal, free lives.

The antibiotics and growth hormones given to cattle can reduce a five-year growing process, to two years or less, which wreaks havoc on the animal's system. Similar things are done with other domesticated animals, including forcing hens to lay many times the normal number of eggs in a given time period until their bodies are ravaged and fail. Some crowded animals bleed from growths and tumors that go untended. So that we can have veal, calves are taken from their mothers shortly after birth. Confused and frightened, they suckle and chew on parts of their crates. These young animals are purposely crippled, being confined to such small areas that they can hardly move a few steps and sometimes can't walk out to the viewing area to be sold. Sometimes they are dragged with their legs splayed out. They are killed very young and tender. Do we really need veal parmesan?

Chickens and turkeys suffer too, as their legs become brittle and sometimes break when they can't support their own unnatural weight. Their beaks are painfully seared off so that they don't peck each other while living in extremely crowded confinement. Pigs, after months of tight and filthy confinement, are sometimes skinned or boiled alive. The pigs—cattle too—can be so crippled that they must be dragged or dropped off trucks before having their throats cut, and if they don't die quickly, then they will slowly bleed to death. Witnessing the unnatural life of a domesticated food animal—the transport, and killing in a slaughter house—would make us reserve the word "humane" for the lion that kills a zebra quickly. There is a reason why factory farming of animals has been hidden from public view.

Natural predator/prey relationships, like the lion killing the zebra, are necessary for nature to function, allowing energy to be passed from feeding level to feeding level and nutrients to be passed back to the soil for more plant life to grow. The prey lives a normal, natural life—and is quickly killed—or if it was sickly is put out of its misery. For people in poor countries who must hunt to eat, the practice is sometimes similar to what we see in nature and is much more humane than CAFO operations. The problem with hunting for food—including fishing—is that worldwide there are too many people doing it, legally and illegally, and so our overpopulation makes what could be a sustainable way of life unsustainable. If everyone hunted wild animals in order to eat, the animals would vanish from the world in a very short time. One of my students once said that if humans were to be truly humane then we would all have to become vegetarians. He was probably correct.

Poaching (hunting or fishing illegally) is another human activity that is doing serious harm around the world. Poaching is done by poor people who eat the animals and/or sell the spoils, and it's done by people who simply want to make money. The sale of bush meat in Africa is a major problem that contributes greatly to the potential extinction of many animals, and as one would expect, the problem is growing year by year due to the rapidly growing human presence. This is especially problematic because more and more African people are dependant on the bush meat that is being wiped out. Elephants, primates, hippos and antelope are just some of the many animals in peril, and keep in mind that HIV and Ebola can be linked to butchering and eating bush meat. In fact, one hypothesis of how HIV originally arose in humans is through this process. Some of this meat finds its way to cities such as New York, London and others, as an exotic taste treat for the "smartly" developed culture. Estimates vary, but it is safe to say that millions of tons of bush meat worldwide are taken legally and illegally ever year.

I think everyone is aware of the trade in ivory from elephant tusks that over the years has killed so many elephants. Items like piano keys, ash trays and other goods have been made from ivory. The trade still goes on, even though it is generally illegal and there are stiff penalties. In 1930 there were approximately five to 10 million African elephants, the exact number is not known. By 1979 there were less than two million; by 1997 there were about 600,000 and today even fewer. Another study found that overall elephant populations have been reduced by more than 90 percent in the last 100 years due to human activities.

Asian elephants are on the endangered list, with (by some estimates) 50,000 or fewer remaining. Habitat destruction, hunting and poaching are the reasons for the elephant decline, as is the case with most species. In Thailand elephant numbers have dropped from an estimated 100,000 a century ago to a low of approximately 6,000. Some of them are tamed for transporting materials such as teak out of the jungles or for plowing fields. Some are used in the tourist industry, doing tricks and giving tourists rides. The abuse to these animals can easily be defined as extremely cruel. Elephants as young as three or four years old—and they can live to 70—are dragged away from their mothers, put in small cages, deprived of sleep, food and water, and beaten for days. The animals are stabbed in their feet with metal nails as an inducement to do what their human trainers demand, such as allowing a human to ride on their backs. This sort of thing goes on in other areas of the world, and worldwide the cases of animal abuse in general are too numerous.

Elephants, the largest land mammals, are considered a keystone species, as they knock down trees and bushes, which gives way to a cycle of new grassland that other animals need, and they forge trails and dig water holes that other animals also use. Elephants are a major source of nutrient-rich manure. It is spread across tremendous areas, allowing plant life to grow and thereby creating not only a food supply for other animals but also habitat. Elephant droppings contain undigested seeds ready to germinate, as well as seeds and nuts that some animals like birds and baboons will eat. As with so many other animals, elephant dung is a major seed-dispersal mechanism and some seeds will not germinate unless passed through the elephant's digestive system. And of course, elephants have been shown to be thinking, feeling, highly intelligent creatures.

To some of us this can be a gross topic, but without animal waste—dung that is (urine too)—a major source of nutrients would end. Actually, animal and plant life would diminish dramatically if it were not for the recycling of animal waste. Again, this is one of the ways that nutrients go back to the earth for a whole new cycle of life to begin. Combine this with the rotting of dead animal and plant material and it accounts for a large part of the biogeochemical cycles that keep life going by recycling nutrients such as nitrogen, phosphorous and carbon.

Dung beetles exist worldwide. They find piles of animal dung and make small balls of it, then roll it onto the forest floor, spreading nutrients and undigested seeds. Some of them bury the dung a few centimeters (about an inch) below the surface, and some of them eat it, which also spreads the nutrients. This is their function in life, and this is not trivial because without these insects, ecosystems would be missing a key component. Dung beetles also dig and bore holes in the soil, which allows water and air infiltration to increase soil fertility for plant life. The holes also help reduce water runoff into lakes and streams, which reduces algal blooms and water contamination from nutrient runoff. This also reduces the numbers of certain flies and parasites, thereby keeping a balance in that department, too. Yes, in nature's eyes dung beetles are more important than people. Sorry about that one.

As mentioned before, about one half of the food we eat is the result of animals pollinating plant life. Put simply, pollination is the moving of pollen from plant to plant so that fertilization and growth takes place.

Insects are the main pollinators, but birds and bats are also very important to the process. Wind also pollinates some species, including corn. The world's food supply may be at risk if the pollinators decline too far.

Bees have been pollinating for 100 million years. They not only pollinate food plants for humans, but they also pollinate wild plant species in another symbiotic relationship that untold numbers of plants and animals rely on. The number of wild plants that particular bees pollinate is incredibly vast, and without question ecosystems would be very negatively affected without these insects. Seventy-five percent of the world's plants depend on pollination, and there are about 250,000 species of flowering plants that need pollination. Two thirds of flowering plants depend on bees and other insects for survival through pollination, but honeybees alone are the primary pollinators of plants that produce food for us, and are responsible for about one third of the food that we eat. Of the human cultivated plants that rely on insect pollination, about 70 percent rely on bees, while the rest is done by flies, wasps, beetles, moths, butterflies, bats and birds.

Wild bee populations are in decline; although it is not known by how much, the decline is thought to be substantial. Habitat is destroyed as human residential, commercial, recreational, agricultural and industrial areas grow and grow all over the U.S. and the world. Logging too, is devastating to bees even if the forests are replanted and farmed because the bees' patterns and habits are so disrupted. Endless development destroys bees' nesting and feeding grounds as it does for other species. Even too many nighttime lights may disrupt some pollinators' functioning.

Widespread pesticide use is killing insects including bees, not just in agriculture but in home use when people use chemicals on a regular basis to unnecessarily kill so many different living things on lawns, shrubs and soils. Bees are attacked with specially designed high-pressure cans that spray out poisons that will wipe out a nest in no time flat. Corporations have convinced the public that insects, not to mention dandelions, crab grass, etc., are bad things that should be killed on sight. Moreover, let's not forget the residual effects of chemical use: poisonings, cancer and sickness in adults, children and animals. An attack on a bee's nest at a backyard party is often a fun and jovial event. But food shortages and starvation are not funny, and that is what will happen if honeybees disappear, especially if the other pollinators continue to diminish. It has been estimated that at the current rate of domesticated honeybee decline, by 2035 there will be no more in the U.S.

Things have gotten worse for the honeybee, as several *New York Times* articles have suggested, including one on April 23, 2007. Additionally, on May 16, 2007, and November 3, 2007, respectively, CNN and PBS reported on the honeybee crisis in the U.S., Canada and the world. As much as 33 percent of domesticated populations have recently and rapidly disappeared in just the U.S. and Canada. It is called colony collapse disorder (CCD). From Europe to South America and in 35 U.S. states, honeybees are vanishing. In some parts of the U.S. there has been an 80 percent decline of domesticated honeybees. Suggested causes range from an AIDS-like virus to a fungus, pesticide use, malnutrition, genetically modified crops, pollution, climate change—or a combination of these factors, such as pesticides weakening the bees' immune systems and allowing a virus to take hold. Bee pollen has been found with many different chemicals in it, including pesticides. France and China have experienced serious declines in pollinators directly caused by pesticide use. Something natural also may be combining with what we do unnaturally, and all pollinators—bees, birds and bats all over the world—have been in decline for decades. Scientists have been warning about this problem for many years.

Honeybee colonies are domesticated and moved by truck all over the U.S. (and the world) to do the job that wild bees cannot do. More and more domesticated honeybees are needed, and we are increasing the stress on honeybees and forcing them to pollinate larger and larger areas as we grow our demand for food. We are forcing them to produce more than is normal. The trucking and workload may also be part of the problem, as they are simply unnaturally stressed out. Fruits, vegetables, nuts, cotton and even forage crops that dairy and meat cattle depend on will vanish without the honeybee. The following is a list of some of the food crops that honeybees pollinate: asparagus, broccoli, Brussels sprouts, carrots, cauliflower, celery, Chinese cabbage, collard greens, cucumbers, dill, eggplant, garlic, kale, kohlrabi, leeks, lima beans, mustard, onions, parsley, peppers, pumpkins, radishes, rutabagas, squash, turnips, beans, cantaloupe and watermelon. There's more. Cotton, flax, safflowers, soybeans, sunflowers, almonds, cashew, chestnuts, coconuts, macadamia nuts, apples, apricots, avocados, blackberries, blueberries, cranberries, gooseberries, raspberries, strawberries, cherries, grapefruit, lemons, mandarin oranges, nectarines, tangerines, kiwi fruit, mangoes, passion fruit, peaches, plums, alfalfa, buckwheat and clover.

The value of U.S. crops dependant on pollination is estimated at around 15 billion dollars, some say double this if we figure in the pollination of crops such as alfalfa, which is used in the livestock trade. One study put the overall value of bees, dung beetles and other insects to the U.S. economy alone at over 55 billion dollars. Dependant on pollination are over 90 crops, and honey bees take care of 80 to 90 percent of the job. To get one liter (a little over a quart) of honey, honeybees must take the nectar from millions of flowers. Honey is nectar that the bees have regurgitated and dehydrated. It is thought that bees communicate by wing-fluttering to pass on information about distance and direction to a nectar source. How amazing are these little creatures that so many of us shun and kill. I'm not saying pet a bee; you might get stung. I am saying that when we put the importance of a particular species in perspective, it should be quite frightening when we look at what we are doing to all of the world's species collectively. Further, even if we solve CCD, we should take this event as a warning of more things to come, and stop taking for granted, creatures large and small.

Bats too, have been under siege for the same reasons that insects and other species are declining. Bat species that eat fruit are of major importance to the dispersal of seeds. Through this process, as well as pollination, they are responsible for the survival of hundreds of plant species. They are key players in the survival of rain-forest ecosystems, and some of the plants they pollinate include bananas, mangoes, breadfruit, cashews, dates, guavas, avocados, cloves, and mahogany. Bats are also very important in keeping down insect populations, including mosquitoes that carry disease, over extensive areas. Yet some species of bats have already been reduced by over 50 percent.

Birds (many of which are pollinators) and countless other animals rely on fruits and seeds as a major part of their diet, foods that will not be there if the pollinators substantially decline. Plant communities that are supported by pollinators bind the soil, retain water, prevent erosion and nutrient loss, keep some water bodies clean, become habitat and food, help maintain climate and produce oxygen. Did I say species depend on each other?

In some forests there are ants that inhabit plants that are then ant defended! The ants live in and on the leaves, and any time another insect lands on a leaf in order to eat it, the ants attack the insect and kill it or drive it off. The plant supplies the ant with a home, and sometimes even a liquidlike food, and the ant protects the plant. Severe damage or even death of the plant takes place when the ants are removed from their symbiotic partners.

Sharks are a keystone species being slaughtered at record rates, and if their numbers drop to a critical point, the oceans will be severely affected. Many shark species are already threatened and may be critically reduced in numbers in 10 or 20 years. Sharks are major predators and of extreme importance to the oceans of the world. They keep other species' populations in check and of course recycle nutrients, all of which maintains food chains. Sharks also have symbiotic relationships with other species—pilot fish and remora fish, for example. One follows the shark, and the other attaches to it. Both rely on sharks for food, generally bits missed by the shark after a kill. The remora actually cleans parasites off the shark's skin (this is common in the animal world). On a larger scale, in the Caribbean it has been shown that coral reefs might be destroyed if sharks disappear. This is in part because the sharks control populations of algae-eating fish that maintain coral reefs. Reefs are home to, and a food supply for, enormous numbers of marine life.

In recent times somewhere around 100 million sharks have been killed in a single year for shark-fin soup, to be used in aphrodisiacs, for their skin, their teeth and bones as ornaments, for oils in cosmetics, and of course, for fun. Many are killed as a result of commercial fishing for other species such as tuna and swordfish. In Japan, shark-fin soup has sold for 150 dollars a bowl, and for some in Japan and China, it's a must-have delicacy. As far as an aphrodisiac is concerned—a lot of animals are killed for this reason—I would suggest sending some Viagra to those who use animal parts to get an erection—free if necessary. We could cover the cost from subsidies to the nuclear and fossil-fuel industries. Animal body parts, such as shark teeth or jaws, as trophies or novelties seem to thrill many people around the world, being another must-have to keep some people happy.

The process of shark finning is truly savage. The shark is caught; its fins are cut off, then the animal is thrown overboard while still alive and bleeding. Whether the shark dies from blood loss or from not being able to swim and pass oxygenated water through its gills is irrelevant. The shark fins are only about 5 percent of the animal's body weight, which shows the incredible waste.

Then of course there is shark fishing for sport—reeling in the big one—which kills so many of these animals, even sometimes when they are thrown back. After the movie *Jaws* there was an explosion of shark fishing that lasted for years. I thought the movie was great, but people must understand the difference between reality and entertainment.

Let's look at another score card. The estimates of shark attacks on people vary, in part because some attacks in remote and undeveloped areas go unreported or undocumented. In recent times a general estimate would be 50 to 75 attacks a year worldwide, with about 15 to 20 deaths. This could be averaged down a bit if we go back through a 100 year period because of the lower human population and therefore fewer people in the water. So in the past 100 years there may have been on average about 50 shark attacks per year. That's about 5,000 over the 100 year span. The attacks are typically about food and mistaken identity; the shark thought the person was a sea lion or other prey. In the recent year in which possibly 75 people were attacked by sharks—and they did not all die—about 100 million sharks were killed by people, which is 1.33 million times the number of people attacked in that year, or 20,000 times the number of people attacked in 100 years. I'd say we won the game! I'm not trivializing a shark attack or demeaning the people or their families who have suffered; I'm just putting things in perspective.

The world's oceans are overfished, and marine life is under assault as humans take out and kill more living things than the system can put back naturally. Global warming is heating the oceans and also changing ocean chemistry and is therefore disrupting the system even further. Pollution and global warming are killing marine life directly, as well as killing and damaging coral reefs—home to over one-quarter of marine species—over vast areas of the planet, damage that also kills countless more marine life by removing their habitat and food. Of the world's fish populations, about 75 percent are severely depleted or overexploited. Tuna, marlin, cod, halibut and swordfish have diminished by up to 90 percent. Bycatch is unintentionally caught marine life. Commercially about 25 percent—about 30 million tons—of all marine life caught is discarded as bycatch and most of these animals die. A recent study showed that in the next 40 or so years, the oceans may no longer support a viable food supply for humans. We are destroying ocean life.

Lowland gorillas are under siege from war, habitat destruction, poaching, disease and the bush meat trade, and by some estimates their numbers have declined by over 60 percent across their range in Africa. However, among the most endangered animals in the world, mountain gorillas are at extreme risk of soon disappearing for the same reasons. They are powerful animals and could easily kill a human, yet they are highly intelligent, gentle and even affectionate, but of course they will fight to the death to protect their young. They are herbivores, like to play and are basically peaceful and only aggressive to humans if we threaten them. We share a common ancestry with them as primates, although we are not evolved from them but rather took a different evolutionary path. As the name suggests, mountain gorillas live at high elevations in mountain rain forests in East Africa. They have very little suitable habitat remaining, and their numbers are now as low as 700 individuals. Some unscrupulous dealers try to capture baby gorillas. As long as adults are alive it is very difficult to take a baby; therefore the adults are shot so the young can be taken. The most appalling example of poaching comes when gorillas are found with their hands and/or heads cut off. That's right: someone out there buys these items for souvenirs. Baby gorillas that were not taken have been found huddled around their bleeding, headless and handless mothers. Over half of all primates face extinction in the near future.

In Rwanda in the 1990s, civil war killed over half a million people and created refugee camps for another 700,000 to 800,000. This cut into gorilla habitat and continues to undermine their existence, and this same environmental damage and extreme habitat destruction kills many other species of plants and animals. In 2006 in Darfur, Sudan, approximately 400,000 people were killed and about two million displaced, while women and children were raped. Our quest to kill each other kills so much else. Extrapolate this out to the endless human wars year after year throughout recorded history. Think of the Vietnam war and the 19 million gallons of chemical defoliant (herbicides)—most of it Agent Orange, containing dioxin—that rained down indiscriminately, generally from aircraft on the jungles of Vietnam, Laos, and Cambodia. Almost two million acres were turned into a wasteland, so that the enemy could not hide or eat. How much animal life did the poisons kill? At the time hideous birth defects in newborns and cancers in civilians and soldiers were seen. The lingering health effects of these chemicals are still seen today in people from both sides of the conflict. Napalm ended a lot of life, too. Now think of the big conflicts like World War I and World War II, in which large areas of the world were devastated.

As with most large predators, bears have little to fear from other animals except the human one. Since the arrival of European settlers in America, grizzly bear (brown bear) numbers have dropped by over 90 percent (other species also declined). In some areas numbers have gone up, but only to what is called a cultural carrying capacity—in other words what human expansion will tolerate, not what nature would have preferred. The spread of cities, housing and condo developments, roads, stores, manufacturing plants, golf courses and agriculture as well as hunting have all contributed to their decline. Bears need large amounts of space to forage and live, and grizzlies may range from 500 to 1,000 square miles. Bear species around the world are declining, and when they come in contact with humans due to our encroachment on their space, sometimes they are killed. They are hunted for trophies, as well as for their gallbladders, which are used in parts of Asia for various medicines and possibly as an aphrodisiac. Any bear will do, so they are sometimes hunted with dogs, scared up a tree or cornered, then shot. The gallbladder is cut out and the carcass is sometimes left to rot. What happens to the bear cubs that might be around? The same thing that happens to most animals under those circumstances; they starve to death or become prey to other animals.

Bears too, are keystone species. As scavengers they keep the forest floor clean, locating carcasses to eat, which disperses nutrients over large areas. When eating vegetation, they become planting machines because their dung is loaded with seeds. They dig up acres of soil in their search for roots and small prey, which leads to new plant growth. And of course, as with any predator, they keep the populations of many plants and animals in check.

By 1975 most U.S. grizzly habitat had been eliminated, and the animals became endangered. Efforts to reintroduce them have helped, but resistance from state governments, ranching and livestock associations and the timber industry have hampered the cause. Once again the demand for development and growth has gotten in the way of another species' existence.

Bears can be dangerous—some species more than others—and we should stay away from them, but again, usually our invasion of their territory leads to the misunderstandings and conflicts; otherwise they generally want nothing to do with us. During the period from 1975 to 2002 something like 62 million people visited Yellowstone National Park, and about 32 people were injured by bears. There have been approximately six known fatalities from bears in Yellowstone. During the last century in North America, a little over 100 people are known to have been killed by bears (black, brown and polar bears). There have undoubtedly been more fatal attacks that were undocumented. There have also been numerous injuries involving bears. However, humans have killed tens of thousands of bears.

Timothy Treadwell, the famous "Grizzly Man" naturalist, spent 13 summers camping in Alaska for months at a time to study grizzly bears. Unfortunately Timothy pushed his luck, and in 2003 both he and his girlfriend were killed and eaten. Yet this event shows that bears are not as dangerous as previously thought. Timothy spent many years closely interacting with bears, including touching them and getting between mother and cubs, yet the bears still accepted him. It wasn't until he came across a bear that some think was sick and couldn't hunt anymore that he finally ran out of luck. It was probably inevitable. Certainly, do not play with bears, ever! The dangerous thing about them as well, as other top predators, is that when they do decide to go after a human, that's it.

Of the five remaining species of rhino in the world—they range from Africa to Asia—all are endangered. Currently black rhino numbers have fallen below 4,000, a 90 percent decline over the past 60 years, and worldwide all rhino populations are down by 90 percent since 1970. They are powerful survivors, having been here about 60 million years, and now they are being wiped out by hunting, poaching, and habitat destruction. Their population has been decimated in recent years even though efforts have been made to stabilize them. One of the reasons they are killed is for their horns, which are prized as trophies. Knife handles are carved from them in the Middle East, and they are also used in Asian medicine and once again, some say, they are another form of aphrodisiac.

Tigers also may be near extinction, and the same reasons run again, with prey loss added to the mix. Tiger populations are down by possibly 95 percent since the beginning of the 20th century, when there were as many as 100,000. There may be only 5,000 to 7,500 remaining in the entire world. For hundreds of years in traditional Chinese medicine many ailments have been and still are considered treatable with tiger parts, so the animals are killed. (The Chinese use hundreds of plant and animal species for medicine, and eat many different animal parts, including tiger penises and paws, all of which seriously adds to species extinction.) Tiger penises are said to be used as an aphrodisiac, although this too has been disputed. Some of the other tiger body parts used are the bones, skin, eyes, brains, whiskers, fat and tail. Is any of this good for what ails you? I don't know, but using tigers this way sure is good for killing off tigers and all the other species that depend on them. I must note that tigers can be very dangerous to humans, especially in places where we have killed off much of their prey and crushed them into smaller areas through habitat destruction. In areas of India, for example, tigers have realized that humans are easy prey and hunt us on a regular basis, killing one or two humans per week.

Even so, think about this: excluding war, and only in modern "civilized times," how many people, including children, are killed every year by other people out of anger or for money? How many others are raped, kidnapped, tortured or generally hurt by other people? The answer is tens of thousands! Humans are much more dangerous to themselves than other species are dangerous to humans.

Dolphins are mammals that have been shown to have incredible intelligence. A bottlenose dolphin's brain is larger than a human brain and in some ways as complex. No one really knows how intelligent these animals are, but some believe dolphins are much smarter than dogs and chimpanzees. There are those who take it even further by comparing dolphins to humans as sentient, thinking, feeling creatures. After spending some time with dolphins, I believe they are all of the above.

During a medical treatment on a dolphin in a tank, another dolphin swam up to this sick dolphin and pulled a hypodermic syringe from it and then showed aggression towards the veterinarian who had inserted the syringe. Fisherman who hauled in nets that had been knot-tied at the bottom so the fish would not drop out routinely found the knots untied and the fish gone. The dolphins did it for a free lunch. Dolphins have saved people from drowning and from sharks, and in the wild they have come up to swimmers and attempted to play. Dolphins have even been used to help people who have serious psychological problems such as depression.

In captivity dolphins have shown increased mortality, some only living half their normal life span. During the capture process many die, although this depends to some degree on who is doing the capturing, from what country and under what regulations. In captivity some dolphins get sick or die from something as simple as not having a large enough enclosure to move around in, or from shallow water causing heat problems, and some calves born in captivity also die. The lack of space does drive some of them into repetitive behavior like we've seen in the polar bear and the wolf. It's called stereotypy. There is even a case of a captive dolphin ramming the side of the concrete tank and dying. The researcher there felt this was a suicide.

Many dolphins are trapped in nets as bycatch of fishing and so drown. About 300,000 cetaceans (dolphins, whales and porpoises) are killed every year from being entangled in fishing gear alone. They are ruthlessly hunted, killed with dynamite, have collisions with boats, suffer habitat destruction and pollution and many die from these things. The Japanese alone, slaughter thousands of dolphins every year. Pollution kills adults and sometimes also the young just because they ingest mothers milk containing dangerous levels of PCBs, heavy metals, radioactive contaminants and other pollutants.

Both dolphins and whales rival a modern submarine's sonar. Whistles, clicks and other bursts of sound allow them to navigate, find prey and communicate. Any excessive noise from boats or even whale watching activities can disturb these creatures. The U.S. Navy is testing Low Frequency Active Sonar (LFAS), and soon this technology may be fully deployed to monitor the world's oceans for submarines. Intense noise is generated in order to bounce sound waves off an object and identify it. The result is that the planet's oceans may soon be completely polluted with noise that can cause dolphins and whales pain, disorient them, and even injure or kill them. Tests of the LFAS systems have paralleled perfectly with numerous strandings and the death of large numbers of whales.

All over the world, many beautiful marine creatures are killed by being entangled in thoughtlessly dumped fishing nets and fishing gear, while others die from ingesting some of the thousands of tons of garbage, plastic debris, and even balloons from celebrations that settle into the ocean.

In the pet trade countless numbers of animals die in the process of trapping and shipping, while many more die in captivity. Lizards, snakes, parrots and others are shipped in secret compartments of crates and the like, out of "exotic" countries and into countries like the United States so that we can have a new and fun pet. In some cases 80 percent of the tropical birds and reptiles that are taken for pets die in transport. Imported into the United States illegally in a given year, collectively, are hundreds of thousands of birds, reptiles, amphibians, mammals and fish. Other countries do similar things.

In areas of the world dogs and cats are eaten and also used for fur, as a December 12, 2005, CNN exposé showed—actually this is very old news long overdue. Many toy animals shipped into the United Sates are made of fur from dogs and cats. In China the number of dogs and cats used for the fur trade may be over two million a year. They are crammed into metal crates and simply thrown off shipping trucks. They are often skinned alive and even hung by the neck until dead. German shepherds and Labrador retrievers are dogs commonly used for fur. Both dogs and cats are housed in tight, filthy conditions where they are perpetually frightened until they are killed while they are conscious, with no painkillers. It is very possible that the toy stuffed animals that children play with, or the fur jacket, collar or fringe that we wear, is labeled as something other than what it is: a dead dog. One estimate revealed that over 25 percent of the stuffed animals out of China are made of dog or cat fur. There is talk of change.

Then there is the commercial fur trade where countless other animals are trapped or raised in captivity in brutal conditions, only to be killed and skinned. The fur trade is global, cruel—and unnecessary. Synthetic coats are just as warm—or even warmer than fur. Worldwide everything from deer, bison, water buffalo, seals, kangaroos, elephants and zebras to crocodiles, lizards and snakes are also killed and skinned for their coverings. (Just about every animal, insect to top predator, has also been used as a food supply.) Nothing has escaped the hands of humans.

Leaf cutter ants move plant debris across the forest floor, spreading nutrients. Worms and other insects tunnel paths through the dirt, helping water and nutrients to get into the ground for plants to grow. Squirrels spread nutrients and seeds so the mighty oak can grow. Birds drop nutrients and seeds in vast, uncountable quantities all around the world. Even salmon eaten by a grizzly bear will have their phosphorus and nitrogen spread inland on the forest floor, not just from the bear's defecation, but from the uneaten fish releasing nutrients into the soil since sometimes bears only eat the most nutritious part of the fish and leave as much as 75 percent behind. Foxes, eagles, insects, birds and other animals depend on the stored energy (lots of protein and fat) contained in these salmon remains. The links among forests, wildlife, and salmon are extremely important. Diminish the bears and salmon, which is happening, and the whole system degrades.

Tropical forests cover less than 10 percent of the earth's surface yet are home to more than half the world's plant and animal species, and they are an important part of the climate system. Yet they are being rapidly destroyed. Each year during the 1990s, 40 acres per minute of tropical forest were lost. Even if these forests aren't completely cut but instead only fragmented into islands made up of forest surrounded by new grazing or farm land, damage is still done. Even if forest is only crisscrossed with roads, this still causes problems for animal and plant life, for reasons similar to those from eliminating surrounding forests, such as animals not being able to cross roads or being unnaturally exposed to predators, as well as the climate being changed.

For instance, what was considered a large block of South American tropical rain forest was left intact while much of the surrounding forest was destroyed and leveled to field. Scientists studied the diversity of life within the island. In the areas around this island-forest, the ground was open and exposed to intense sunlight. The ground became eroded and scorched, losing its nutrients and ability to grow native species. The landscape surrounding the island was changed to such a degree that the species of plant and animal life that had previously been in the forest probably could never come back, and it became questionable whether the island-forest could survive in the long run. This was in part due to the climate of the area being changed because of tree loss. As with other plants, as part of photosynthesis, trees give off very large quantities of water through transpiration. The same pores (stomata) on a leaf that take in carbon dioxide and give off oxygen as a byproduct of photosynthesis also give off water vapor as a byproduct. This moisture contributes dramatically to cloud formation and rainfall patterns, so much so that when enough trees disappear, the climate of an area can be permanently altered. A tropical rain forest may become subtropical or desert-like. Get rid of enough trees, and the self-replicating climate may change to one that will not support life the way that it previously did.

Predators that normally couldn't easily get to animals within the forest now could find their way in through the edges of the now less-deep forest. New prey for these predators meant fewer of the animals natural to the island. In addition, trees near the new forest edge were suddenly exposed to light and dry wind and so were decaying and dying. This caused the die off of many insects, as well as the birds that relied on insects for food. Spider monkeys started starving to death as the amount of remaining forest was not enough for them to find food. They are tree dwellers and therefore could not cross open ground to try to make it to the next patch of forest. They vanished. No monkeys meant that certain fruits went uneaten and so rotted without their seeds being spread throughout the forest, which led to less replacement growth of fruit trees.

Wild pigs in the forest had a habit of rolling on the ground in puddles, which carved them out into slightly larger puddles, called wallows, that would further fill with water. Monkey frogs relied on these puddles to survive. Their eggs would hatch in the trees, and the newborns would drop into the puddles to continue developing. This was part of their life cycle. Because of the diminished forest size, the pigs starved to death because they didn't have enough forest left to find food. The result was no more puddles, and so the frogs died off. It is likely that the forest and its animals will change further as other symbiotic processes continue to be altered.

I have used as examples only a handful of species and how interdependent they are on each other, in order to make more apparent how hard and fast the dominoes can fall when humans disrupt ecosystems. The interdependency of the approximately 1.8 million known species, as well as the countless unknown species, cannot be emphasized enough. A massive, worldwide collapse of ecosystems brought on by the out of control extinction of species is starting. Although we don't know exactly what will happen to humans if we kill off very large numbers of plants and animals, it should be obvious that we will be affected.

More and more people who work with animals, from dogs and horses to bears and elephants, are convinced that they think, feel, and have emotions. Chimpanzees and dolphins—already mentioned—have been shown to be self-aware and to have complex emotions and much higher intelligence than previously thought. Where we draw the line is hard to say. I don't think of an insect the same way that I think of my dog that I love. However, a worm or an ant deserves no less respect than a dog or dolphin because of what they do for us. Just look at the pollinators. Should I crush the bee to the ground with my foot just because I can? Many people are repulsed by creatures that don't fit the mold of what we see as furry and cute. Yet we even kill the furry, cute ones. There's a giant disconnect between us and them.

When an animal kills, it does so mostly for food, and occasionally over hierarchy—humans do this too. But animals do not kill tens of millions of their own in wars over religion, cultural differences and greed. They do not damage or destroy everything around them that keeps them alive. If they are not capable of these things because they do not possess the minds that we do, then we should be able to learn something from them. There are no vicious animals, except one lone species.

Animals run on Darwinian fitness, whereas we run more on human fitness, and we obviously should not go back to pure Darwinian fitness. From nature's standpoint the man in the wheel chair should not be here, but from our standpoint he should. He still can contribute something to our society, be it something small or large, like the mind of a great physicist. An individual can make a contribution to a species such as ours without having big claws, strong legs and healthy genes to pass on. However, we cannot sustain continued destruction to the Darwinian world in order to expand our human world. We must find our balance within the world of plants and animals. That means using our intelligence and technology to reinject Darwinian reality into our veins in a civilized and careful manner. Population control, conservation, education, proper technology use and respect for "them" would do just fine.

I have stood in my vegetable garden, inches away from a bee that was traveling from bright yellow flower to bright yellow flower on a cucumber vine, pollinating for me. I am in awe of the process and am thankful to the bees. I never get tired of watching them work so hard for us, for free. I wouldn't think of killing one of them. I might run like hell if I thought the little thing was after me, but I still wouldn't kill it. Are there times when we need to kill animals? Yes, of course. If a swarm of bees was attacking a person, or if a wild animal was doing the same, then we would have to defend ourselves. But the conflicts between people and other animals that are life threatening to us are very rare. As we have seen, more commonly it is the other way around.

It is also common to hear people refer to a person as an animal, when that person has done something that is considered disgusting or heinous. Saying that such a person "is an animal" or "behaves like an animal" is part of our cultural slang. However, the saying doesn't make sense because animals' actions do not warrant the negative connotation.

We threaten them and their very existence, and so our existence is in peril. The current level of species extinction may quietly get us in the near future, while we sleep in ignorance of it. Without us, they would do just fine. Without them, there is no us!

9
ECONOMICS, THE LIE

The world economy has expanded by approximately seven times what it was in 1950. Adjusting for purchasing power differences between countries, gross world product (GWP) in 2007 was about 60 trillion dollars. On a price-adjusted basis, GWP has grown over the last century, by approximately 18-fold.

Commentators and business leaders cheer when world economies continue to grow rapidly and dramatically. On the typical television financial or news program, they claim everyone should continually expand economies and boost gross domestic product (GDP). Will we double the world economy from here, again and again? Shall we go from 800 million cars in the world to two billion, four billion, then eight billion or more? Should the United States go from its approximately 220 million cars to 440 million and eventually a billion cars? Estimates are that the number of vehicles in the world may be over two billion by 2050 as China, India and other developing nations continue their upward spiral of rapid economic development and while the developed countries like the U.S. keep growing their economies.

The financial world says U.S. builders should start over 160,000 new home units every month, forever. Jobs need to be created for the growing population, and if we average 150,000 new jobs every month forever, that's great, but more would be better. Can we cover every square foot of the country, and the world, with houses, roads and commercial enterprises, and so kill all the pollinators? Will everything around us, all production and consumption, endlessly magnify?

Politicians and business leaders say we must keep expanding economies in order to increase government revenues such as income tax, social security, property taxes and sales tax, so that services to the growing public can be maintained. However, as population increases so does demand for public goods and services—resources, fire, water, sewer, garbage, schools, infrastructure maintenance, etc.—which contributes to driving up the cost of living. If government tax revenues do not keep pace with rising costs, or if revenue decreases for some reason, such as too many poor people, too many people not paying into the system (the U.S.—along with other countries—has a gigantic underground cash economy), or the declining middleclass, then more money is needed.

It is becoming increasingly common in many U.S. cities and suburban areas for more than one family to live in single-family dwellings. Often, rent from the additional family helps pay the bills, and millions of people are living this way. The demands this puts on an area exceed what the area was originally designed for. Increased population density, however it comes about, contributes to the spiraling up of the cost of real estate, property taxes, and the cost of living in general. These costs have risen all over the United States in recent years as population has increased to record levels.

There is a never-ending cycle of more people and more growth to fix many of the problems, created by too many people in the first place. It's a giant Catch-22. We are perpetually trying to outrun ourselves as though there were no upper limit to our velocity. In reality, what our leaders are saying is that our economy, our way of life, can't continue without increasing consumption to boost GDP, and that increase in consumption is driven to a great degree by population growth. This means that in their minds, there should be no limit to population growth, and therefore a discussion of carrying capacity is out of the question. Someone had better remind the people in charge about the limits of food, cover, space, matter and energy: simply put, the limits of physics and nature.

It has been calculated that if all of today's 6.7 billion people lived in the same style as we do in the United States, we would need four more planet earths to support humankind. The last time I checked inventory, we don't have even one-half of an additional planet for backup use. World population and consumption is growing, while increasing consumption is encouraged, and too much of the world wants to live in a manner similar to the United States and other developed nations. Capitalism and the economies of the world in their present forms are not sustainable.

There is no question that democracy, capitalism and free trade tend to bring nations together and even contribute to peace as the world's citizens have a free and better life. Japan and the U.S., two countries that once fought to the death, are now partners on the world arena. China and India have become capitalistic buddies with the U.S., and they are reveling in the relationship. However, this is too much of a good thing, as we are trying to make the whole world one big consumption machine.

When ecosystems collapse, economies will come down hard, and the standard of living for all people, rich and poor, will come down while the quality of life degrades. Nations will be at each other's throats. It is impossible for the developing nations to ever fully catch up to the developed nations in standard of living and consumption levels. Their attempt to do so may in the end only bring more wars as competition for food, cover and space heats up. At the very least starvation, genocide, suffering, local wars and continued political strife will intensify. Sorry, but as the economy (in its present form) gets "better", the environment degrades, because the currency of the economy is the environment—that is the air, water, climate, resources and the plants and animals that keep us all alive. The economy works because of the environment, not in spite of it.

Most economists, business leaders and politicians have a common goal in mind, which is to get every American, and everyone else in the world, to consume as much goods and services as their earnings and credit will allow. Commercialism has become extreme. Excessively long and numerous television and radio commercials, print ads, internet pop-ups, junk mail, junk faxes, junk e-mail, billboards and signs and lights abound, all in a nonstop barrage to get people to buy and consume as much as possible as quickly as possible. We can't watch a television program without pop-up ads regularly coming across the screen and distracting the viewer with some kind of reminder or advertisement. It's as though the people running the show don't want to entertain or educate as much as they want to brainwash us into buying or recognizing their brand. The subliminal imprinting aspect of all this advertising is extreme, and I believe that this is debilitating to the human mind and its ability to function clearly, especially for children.

Our eyes and ears are subjected to a rapid bombardment of images that relate to all our physical and emotional senses, stimulating them, and temping us to eat, drink and buy during every spare moment of our lives. Marketing has reached intense levels as it attacks children and adults emotionally and sexually to prod us to buy, buy, buy. People's self-worth and very psyche too often are defined by money and material items. What to own, wear, drive, smell like and look like are predefined until children growing up can no longer think for themselves. Their thought processes lose depth while their ability to think critically fades, submerged in a world of superfluous, meaningless things. Environmental issues, conservation and a more subdued, sustainable way of life become contrary to what we believe matters. Many times happiness for our children and ourselves is not there unless we can buy and consume, so happiness ends up as an illusion.

The food industry has done a great job of constant and intense advertising even as Americans have the distinction of such prevalent weight and obesity problems that many hospitals stays, as well as sicknesses and premature deaths are related to our eating habits. In March 2006, the U.S. Surgeon General stated that obesity is more dangerous than terrorism. The attacks on September 11th killed 3,000 people, whereas over 250,000 people die from weight-related ailments every year in the U.S. But if we overeat it's good for business, so of course, we eat on just like the food industry wants us to do—no matter what it is that we eat. The cigarette companies and fossil-fuel companies want us to consume large quantities of their products, too. We live in a culture that has taught us that more and bigger is better, and we too often follow the leaders.

Shopping networks and infomercials have become consumption powerhouses, as people sit home and order through the mail, shopping having become an addiction for many. The packaging that manufacturers encase their goods in is often oversized and therefore extremely wasteful of materials, usually plastics and paper, and so there goes oil and more trees. The message is produce and consume more this year than last year, and so on. In other words, create a J-curve for everything from energy and resource use, to population and therefore environmental damage, because its all good for the economy.

Popular financial programs on television, routinely interview CEOs and other executives of large and small corporations. Every corporate leader says that their business will expand and grow every year, forever. More stuff will be sold next year than this year. All over the world factories are going up. Houses, trucks, cars, jet aircraft, air conditioners, computers, toys and everything possible will be sold to or used by everyone on the planet. Every politician and business leader promises this continued economic growth, with little idea of what it really means to the planet and our existence.

Every financial analyst tells us that the best place for our money is in the stock market because history shows that it always goes up in the long run, and they are correct. Look at Figures #1 and # 5. We've already seen the J-curve in Figure 1; now look at the J-curve in Figure 5, which represents the Dow Jones Industrial Stock Average over the past 100 years, a measure of the U.S. economy and sometimes that of other parts of the world's as well. Notice the Dow curve is following the population curve all the way up. If the two charts are overlapped and compared, we see the explosive growth in world population and the DOW start at about the same time. As with all other blips down, the drop in prices around the year 2000 certainly was not due to population drops but rather to temporary factors as the economy caught its breath. In 2008 we had a dramatic downside correction, caused by a credit crisis resulting mostly from corruption.

When people say that stocks will always go up, what they are unknowingly saying is that food, cover and space are unlimited, and that the world is an infinite place with unlimited resources. Both J-curves on the charts imply that the human species can grow forever. The main reason why the Dow keeps going up and likely will for some time is because production and consumption in the world are increasing to unprecedented levels because of unprecedented population growth and consumption, creating more demand for goods and services, which in turn increases profits for the corporate world. This is a phenomenon that will be short lived because we are living in an approximately 100-year span when a key reason why the current form of capitalism has been working is because this extreme consumption hasn't collapsed the earth's ecosystems—yet. The earth has been capable of absorbing the damage over the past 100 years (more if we go back to the industrial revolution) while giving some of us all that we need and want, but now the earth and its species are approaching the end of their ability to support us. Therefore our population and economy will also reach a turning point.

Look at the other graphs that represent environmental damage caused by greenhouse gases and temperature rise (Figures #6 & #7) and you will notice that they all represent J-curves. If we look at other graphs that represent environmental problems like resource depletion, species death, and new and reemerging diseases, we would see more curves similar to the population and Dow J-curves. Global warming is happening much faster than previously thought. Will melting polar ice and glaciers, sea-level rise, heat waves, droughts, floods, food shortages and fires all start to look like overlapping J-curves? What are the surprises in store for us?

Destruction of ecosystems is the direct result of overconsumption and overpopulation because of the way our economies currently function. Yet there is a major difference between the J-curves of our population growth and the Dow's price increase, and the measures of planet degradation from something like greenhouse gases, indicating global warming. The world population can easily crash, and then so would the Dow and the economy, because production and consumption would reduce quickly and dramatically because there would be many fewer consumers. All the manufacturing capacity that previously throve on excessive consumption would have nothing to do.

However, the amount of damage to the earth at this point would not rely on the precariously existing biological entities called human beings, so the J-curves representing this damage would not crash just because the human population and the Dow crashed. The earth's damage levels represent more permanent change done to the physical and chemical properties of the atmosphere, oceans and terrestrial habitat, and this damage can remain for hundreds, if not thousands, of years. This damage does not reverse overnight just because we stopped adding to it by a forced reduction in both the population and the economy, so a full human comeback—at least in the near term—becomes very difficult, if not impossible.

Let's say for a moment that we stop population growth at the current 6.7 billion people, what then do we do for expansion and growth of our economies? The answer is bring more people into the modern consuming lifestyle. Then what do we do when we have maximized that avenue? At that point, with expansion and growth stopped, the earth would be suffering a constant rate of damage versus the previous increasing one, but that constant rate of damage would be much less sustainable. Remember that as of 2008 only about 1.25 billion people are living well as modern consumers, and now I am talking about over another five billion humans becoming modern consumers. It's already quite clear that the earth can't sustain such a high level of consumption and consequent damage. Yet we are still increasing the population level and simultaneously attempting to get as many people as possible to become part of the constantly consuming masses.

Inevitably the growth and expansion of world economies will stop. This may happen because nature hits us with something dramatic to reduce our population, such as disease or starvation due to environmental issues. It is also possible that we may decide to regulate population growth ourselves because we finally realize we just can't support it any more. It is also possible that population growth will naturally stop and even reverse as cultures naturally change. Clearly a severe and rapid drop in population would crash our economy. However, if we institute a two-child policy, stop immigration and let death rates catch up to birth rates, then our population will slowly and comfortably move backwards and so will GDP. But the changes will be more manageable and we will be able to deal with them and adjust through changes in economic policy and procedures. The classic recession scenario may simply not be there anymore. Of course, I am not implying that there won't be economic problems. Any economy, an old one or new one, will never run perfectly, but we must make the change.

We must carefully develop a new "no growth" economy, one that includes environmental logic and awareness, an economy that runs more on need and less on the relentless drive to grow. The behavior of the stock and bond markets would certainly change, and so would methods of investing in a very different financial world. Further, the developing nations need to come up to some level of humane and comfortable existence, which means as we decelerate the growth, they would still accelerate theirs to some level. However, there is a level of balance and fairness where we all need to be. And, yes, I know this is a scenario that would be very difficult to impose worldwide, but by the U.S. setting an environmental example things might start to change. (And, yes, I also know that zero growth will likely never take place unless disaster forces it on us.)

The U.S. which has built such a gigantic economy dependant on massive consumption and waste to stay economically healthy, will take the biggest fall in any type of environmental collapse and therefore economic collapse. Countries in Africa and elsewhere where people are used to living with much less may take the shortest economic fall. However, as the environment initially starts to change, these countries may suffer the greatest problems with things like food and water.

What does sustainable growth mean? The term is used all the time.

It means that we can still grow economies if we conserve energy, recycle, and generally treat the environment much better than we have so far. However, sustainable growth still insinuates that we can keep growing. No matter how efficient we become and no matter how careful we are with the environment, there will always be limits. Being environmentally aware and doing good things for the environment so that we can continually grow economies in a cleaner, safer manner only extends the day when we must stop growing. Sustainable growth is unsustainable.

Worldwide, economies and population must go backwards in size, not forward. The idea of boosting GDP and growing economies to some extent comes from the ideology that we all should be able to get rich, and then of course down the road the rich get richer as production and consumption increase. The relentless drive to boost GDP is not needed because that drive also comes about to support a growing population, people that represent overpopulation—remember replacement level. Boosting GDP and creating jobs for a growing population is a self perpetuating machine that infers perpetual motion is possible. In reality, growing and expanding economies is to some degree a greed-based concept (much more so in a fully developed nation than in a poor one). That is very hard to say, and no less thorny to read. But we need a different concept; let's call it CCP, for carrying capacity product. In that scenario population is lower and stabilized, and we produce and consume approximately the same amount of goods and services every year at levels and in ways that are sustainable (using solar power, etc.).

The resources and space do not exist for our way of life to grow forever, and even though large quantities of coal, oil and natural gas do exist, the remaining resources for the most part need to be left in the ground. What does it matter how much remains? It's dangerous material to use. To contemplate continued fossil-fuel burning and sustaining additional climate ruin is not rational. Nor is it rational to consider cutting down one more tree, overfishing the oceans for five more minutes, letting anymore soil erode away, killing one more species or damaging the environment in any way, any more. Our economy, the phenomenon that built this great country, is the very thing that will destroy this country and the world if we keep maintaining business as usual. Our system of economics is much like an internal combustion engine in that it is obsolete and must change. The idea continues to be "Consume, consume,

consume. It's the economy, stupid!"

Corporate America has the power to help dramatically with all kinds of new technology, but only if the corporate and political leaders get in touch with reality and become scientifically literate when it comes to the earth. They must stop—we all must stop—thinking of profit only for the sake of a bigger house, bigger car, and more stuff, and instead think of profit more in terms of the existence of the planet. For instance, there is a gigantic monetary benefit for the corporations developing and selling solar energy and other environmentally positive technologies. The amount of money saved by companies that use solar power and conserve energy and materials is substantial and goes straight to the bottom line. However, the business community must understand that environmental improvements go hand in hand with a declining population and an economy that stops growing. The current thought process is unfortunately, that any reduction in pollution, materials and energy use, is simply a way to allow for more growth and expansion of business and economies.

Many corporations have at least figured out the part about profit margins, and the connection between profit and being, to some degree—and I say this with reservations—environmentally conscious. Billions of corporate dollars have been saved through conservation and efficiency, and knowingly or unknowingly, and with caring or not caring, corporations that conserve have reduced pollution and contributions to global warming. Many companies are finding a serious place for solar power. Federal Express, for example, has covered the roof of its main facility with PVs. Whole Foods in Ridgewood, N.J., has done something similar. I hope that somewhere in corporate America's thought process is a desire to better the planet and all its species and not merely a desire for more and more growth for more and more profit. Profit without a conscience is killing us.

When the corporate accountant does the books, he or she tallies up income and expenditures and figures the year's profit. When inventory drops, more has to be purchased or manufactured directly, and expenses will increase. There is an assumption that inventory will always be available, even though its cost may fluctuate. There is a major problem with the way corporations do their books because they do not understand, or do not care, what the real cost of doing business is.

Corporate bookkeepers ignore the earth's inventory of goods and services (that is, how long the things such as livable climate, ecosystems, oil, coal, food, water, soil, etc., will last). They don't figure in the cost of human-induced damage to ecosystem goods and services such as climate, soil, water, wetlands filtering pollutants and protecting coastal areas from hurricanes, and animals and plants generating oxygen and food. They don't figure in the health cost of breathing polluted air and drinking polluted water, or the millions of people who die prematurely every year from environmental-related sickness such as cancer and heart attacks. They don't figure in the dead trees, crops, fish and animals and overall ecosystem damage from pollution, acid rain, global warming and ozone breakdown, all of which add up to billions of dollars lost. If the true cost of doing business was factored in, a lot of companies—especially car, mining, coal and oil companies—would show not a profit, but instead a loss. College courses and books on environmental economics have examined this reality. It's very easy for the corporations to pass these enormous expenditures on to the public, which pays for them in the form of health costs, food and energy costs, and higher taxes for services such as waste cleanup.

Recently a team of scientists and economists got together and put a dollar value on the earth's annual ecosystem service to human beings. It turned out to be a staggering 41 trillion dollars (in 2004 dollars). Whatever the margin of error, why should we not think of these systems as inventory to be drawn from carefully because human existence depends on them? The true value of a piece of land is not how many houses can be built on it, or what kind of condos and golf courses can be set up there, but instead what nature keeps stored there, such as the soil, water, plants, atmospheric oxygen, food, medicines, animals and other resources that keep us all walking upright.

A person can be president, senator, congressperson, Fed chairman or leader of the world's largest corporation and only have to take one or two science classes during his or her entire college life, courses that don't have to be in earth science. Then that person can make decisions every day that affect the planet about which he or she might know very little, if anything at all. Because a leader may have science advisors does not mean that person has sufficient ability to assimilate the data, especially if the advice and data are corrupt.

We are living in an era of unbelievable scientific advancements, and at the same time science and math illiteracy in the United States is epidemic. There have been many articles in newspapers and magazines, as well as television discussions, about this serious problem among our college students, a problem that has gotten worse over the past 20 years. A cognitive elite understands the science and math, and much of this knowledge is profit based, being put to use in making computer chips or fancier SUVs. Earth-science-based knowledge is reserved for a smaller group, almost a fringe group.

The general public's understanding of the earth is minimal, and our educational system does almost nothing to help. This lack of understanding is an epidemic at all levels and has become an accepted way of life. Many very educated people, even those with master's degrees and PhDs know little or nothing about the science of the earth and often look at the world through eyes that gauge existence through their own subject matter, with little or no understanding of the physical limits of this planet.

Election time makes the point. Listen to the politicians speak at debates, on television commercials, on talk shows and at conventions. Almost nothing important or substantive is mentioned regarding the environment. In 2007 and 2008, at some presidential campaigns, candidates mentioned nothing about any issue pertaining to the environment, even with ongoing droughts, honey bee collapse, and global warming accelerating. When climate change was mentioned, the talk was so minimal, if you sneezed you could miss it. The talk is of boosting GDP, producing more and consuming more, creating more jobs for a growing population, building more homes and buying and selling more stuff. Education, the military, social security and medical issues also come up—and all these subjects are very important. Sometimes the talk is uplifting and wonderful, and sometimes it is not. Either way, almost everything said is merely rhetoric because without an understanding of the earth and its support systems, all the good intentions of the politicians cannot materialize. Their promises evaporate if we destroy what supports us by trying to make too many well-intended things happen.

Everyone can't have two cars in every garage and a chicken in every pot. Every citizen can't have a big house, two or three cars or SUVs, food, clothing, education, social security, vacations and all kinds of toys if there are too many of us. As our leaders strive to help people to have a better life and simultaneously boost the economy, they do so under a veil of incredible ignorance. They do not understand carrying capacity and the limits that nature has in place.

The major political issue is generally the economy, and of course it is vitally important. Environmental issues are on occasion dropped in here and there during campaign speeches, but these issues are often reduced to local ones like clean water acts and clean air acts—both very important, but neither of which address the big picture. Today, global warming is finally being discussed, but it is still talked about with minimal understanding or in many cases ignoring the true realities of what must be done to stop the process. A few politicians might understand the dilemma, but they are faced with the conflict between following their conscience and being political and following the corporations, lobbyists and PAC money and not upsetting the public. Politicians know that what the public doesn't understand, the public won't sacrifice for.

There simply is no requirement, desire, or cultural norm that dictates that we must understand anything serious about this planet and how it operates. We are raised with minimal knowledge of the earth and because of this, we subconsciously assume that the good life is some free and limitless economic realm to constantly aspire to. Educational television programs that deal with animals and the earth have too often generated mostly entertainment value, and although some have wonderfully explained environmental problems, there has not been enough urgency placed on the threats that we face. Fortunately, this has been changing in recent times, and there is a collective effort on the part of a few to educate us all about the environment, but this education is not working on a large enough scale. Our children still have a view that success comes from excess growth and consumption. Excess is not even seen as excess; it's seen as normal.

We've come to believe that all is well if all is economically well. The air looks clear, so all must be well. There is no immediate tangible difference in my life, so all must be well. The same physics and chemistry that give us our food, car, perfume, medicine, surgery and everything else we want and need are the same physics and chemistry that are telling us that the planet is at the end of its rope. So far, though, society mostly wants to hear about the physics and chemistry that supply the material stuff. This is changing, but it needs a big push.

One of the most easily obtainable types of information is about economics. The topic is all over all the media, especially television. Financial shows abound to give advice, explain economic topics, whether about GDP, the stock market, retirement or mortgages. As a society we are raised on this type of information and have come to firmly believe that economics and the stock market run the world and all its systems. On the other side of the fence, earth science is not mainstream, and in that field knowledge is more difficult to obtain. Not being part of popular culture, earth science is deemed irrelevant by default. Our society has no serious interest in teaching it, or explaining it to anyone, other than in selected areas where the listeners choose to listen.

When I was in college studying geology, one of our professors asked the business department to have their students sit in on a lecture about the environment. They refused, saying they had no interest. At another college I took some economic classes to be well rounded, and one of the classes was taught by a famous economist who had written many books. I read one of them, on the history of the stock market, and it was excellent, yet there was another side to my professor. He used to get up in front of the class and tell us that overpopulation was a ridiculous concept and that global warming and the like were a farce. I asked him what his background in science was and he proudly replied, "none." I've had more than one economics or marketing professor tell my students that "those scientists" don't know what they are talking about. It is engrained in our culture to listen to, believe in and bow to the people who talk about money.

Ignorance of earth science exists within the U.S. government at all levels, although it is much worse at the federal level and at least improving at the state level. The dichotomy at the federal level always astounds me. On one hand the federal government talks about clean air and clean water acts, and regulations on chemicals, hazardous waste, landfills, etc. (although they don't always do what needs to be done here). On the other hand, the federal government continues to push for increased production and consumption, and so increases pollution, waste, and environmental damage. So far the federal government has done little to propagate serious conservation in the public or within the corporate culture. Regulations are passed to clean up what comes out of the tailpipe of cars, trucks and smokestacks while greater quantities of emissions are simultaneously encouraged. Meantime, carrying capacity and limits of the planet's ecosystems are ignored.

Obviously, the idea of purposely slowing the economy is completely contrary to our system of economics, and if it were done too fast or without very careful planning, the economic results would be catastrophic. Social security is one example where serious problems would arise because it is dependant on ever more workers paying into the system. Starting an additional sales tax might solve this problem, especially considering that many people often don't pay any taxes at all, as they work in the underground cash economy. We need the brightest economists to slowly and carefully reshape our economy. As population growth is slowed and reversed, GDP will moves backwards to some stabilization point about which it will fluctuate, in a restructured, environmentally aware system. This complete economic redesign should be part of our new Manhattan Project. As already mentioned, methods of investing money would change, and the rich might not get richer—in fact, I might not get rich—but so what. We all need to slow down and appreciate the natural world that we are part of, that keeps us alive.

The idea that the law of supply and demand will fix environmental problems is flawed. True, if we raise the price of gasoline, say with a dollar tax per gallon, or if the price rises as demand rises on a limited supply, people will drive more efficient cars and less gas will be burned. A similar scenario is true when the price of anything goes up or down. For many years Europe's substantial tax per gallon of gasoline has worked well at keeping people in small cars and keeping air pollution lower than it would have been, but they still have too many cars and too much air pollution, and we still have global warming.

By subscribing to the idea that we will save the environment as efficiency is forced into play by the economics of supply and demand, we are saying that the physics and chemistry of the planet's atmosphere, hydrosphere (the oceans and water systems) and ecosystems, will adjust in order to suit the cultural, emotional, and economic desires and whims of human beings. In other words, we believe economics will determine what the earth can—or can't—take.

The amount of physical and chemical change that will critically damage the earth is a fixed amount already determined by the laws of physics and nature. This amount of change in many ways is an absolute quantity of carbon dioxide, methane, chemical poisons, species extinction and other factors that, when reached, will put us in the "too late" mode for our survival. The real issue is whether our economy of supply and demand can make something like gasoline, or any other resource, so expensive that the amounts we use will diminish far enough, in time, to not surpass the level of negative change that will do irreversible, detrimental harm to us.

By the point in time when fossil fuels get too expensive to use in large quantities anymore, for example, we may have already passed the point of no return because of the damage that has already been done. If price issues stop the damage in time, fine, wonderful, and thank you! If they do not, then it's too late and that's that. And guess what? The recent price rise in oil and therefore gasoline, fuel oil, home-heating oil, natural gas and other fossil-fuel related products has not come in time. Global warming is here and getting worse, and meantime we are nowhere near getting off fossil fuels. The supply and demand thing didn't work, because we can't dictate to the earth's systems.

Paradoxically, if the price of fossil fuel drops for any reason, such as a slowing economy or a temporary glut in oil—which has happened—then the public tends to increase their fossil fuel use once again. The implementation of energy-saving technology slows and planetary damage increases. We may certainly lessen environmental damage if prices go up in, say, gasoline; however, the lack of understanding of the planet still exists, and so the problem may not be solved. Many other issues don't get fixed either, because we are again waiting for supply and demand to change some other consumption habit.

The serious acceptance of an environmental problem by the political and business community must take place first, otherwise getting away from something like burning fossil fuel takes far too long. If something is profitable and works economically, then change is fought. Traditionally the political and business communities simply don't want to believe what science has to say about the environment. Any such belief interferes with what they have been taught and might cut into money earned. So here we are today. And it's too late because global warming and climate change is upon us. However, is it too late to avoid catastrophe from global warming, or is there still time? We don't know for sure, but if we don't immediately start to make serious changes in the way we run the show, we may very soon find out how much time was left because if there is any time, it may be very short.

Additionally, the economic system that was going to save us has a little thing called profit motive built in. The profit motive for the oil, coal, mining and car companies to keep the world burning fossil fuels is enormous. They can manipulate prices as well as make sure that alternative energy sources such as solar power are delayed in coming to market. If the only energy source out there to heat our homes or to get us to work is fossil fuel, then we must burn it at any cost. We simply will have to spend less money somewhere else. In the past, many corporations have done everything in their power to keep us from getting off fossil fuels, including denying the effects of burning the stuff, as well as lobbying our elected officials for support.

Our societies have become more efficient, but over time we have dramatically increased consumption, and we keep encouraging even more consumption, all of which minimizes any gains in efficiency. The critical link between our system of economics and the earth's physical systems is that no matter what levels of consumption we engage in, damage will always be done. The earth can handle some damage, not unlimited damage.

No matter how inexpensive a resource, the resource must always be carefully conserved. No matter what the price of a resource, just because we can more than afford it monetarily is no reason to not seriously conserve it. In the U.S., the idea of paying a higher fuel price through a tax to lower consumption is fought tooth and nail as consumers demand cheap gasoline. If consumers can afford something, such as gasoline, then they feel good economically, and don't worry so much about how much gasoline they use. However, can our children's lungs or the planet's life-support systems afford the gasoline? Because we are planning carefully for our child's future does not mean that he or she has one. If people say, "not in my lifetime" when responding to environmental damage, what do they mean? Do they care about their own species and its future, or are they making a joke?

Everyone does not have to major in complex sciences, but everyone should be required to take courses in basic, conceptual earth sciences from the time they are six years old throughout their entire schooling. No one should be allowed into any government office, especially Congress, Senate, or the presidency, without adding at least one year of environmental-science studies to their college degree.

When a kid graduates high school or college, he or she should automatically be concerned about stepping on the insect that helps keep soils functioning and should already be aware of what it means to have more than two children. Being cool should not require a fast car or huge SUV; being cool should mean owning a hybrid or solar-powered car regardless of what they look like. Sports and entertainment stars are cool, but so are scientists like Sherwood Rowland and Mario Molina. Maybe they saved the world when they discovered the destruction of ozone in the stratosphere. Why isn't a national holiday in the U.S. named after them?

Many people's priorities too often revolve around what we can drive or eat, how big our house is, how much money we have, how much stuff we can buy and consume and what we own. Call it ego, vanity, materialism or consumption, but the earth doesn't like that aspect of our character. To say that as humans we have the right to do what we do is to say that we have some ordained right to violate nature, to take from everything on the planet and put back nothing, without regard for anything but ourselves, which in reality is like saying "let's commit slow suicide." There is a major cultural flaw in the United States that paradoxically is seen as wonderfully righteous: it is the value put on consuming as much as we want whenever we want, and belief in consumption as a value is amplified by a system of economics that defines this consumption as the measure of success and happiness. We need a complete cultural change in the United States and in the world. Less is more, and smaller is better must become the mindset. The environment cannot be made to conform to our economy; instead our economy must conform to the environment.

10
GLOBAL WARMING AND CLIMATE CHANGE

In 2001 approximately 2,000 scientists from over 100 countries representing the Intergovernmental Panel on Climate Change (IPCC), after analyzing about 3,000 scientific studies showing changes in over 400 biological and physical systems, concluded that global warming is real and that human beings are contributors to the problem. Since then the IPCC has become more resolute, stating in February 2007 that the existence of global warming and climate change is unequivocal.

A new IPCC report in April 2007 says, "A global assessment of data since 1970 has shown it is likely that anthropogenic (human caused) warming has had a discernible influence on many physical and biological systems." The report goes on to say that the resilience of many ecosystems is likely to be exceeded in this century by an unprecedented combination of climate change, associated disturbances (e.g., flooding, drought, wildfire, insects, ocean acidification) and other global change drivers (e.g., land use change, pollution, overexploitation of resources). The report also states that some large-scale climate events have the potential to cause very large impacts, especially after the 21st century. Then in November 2007 another IPCC report became more dire, talking about even more disastrous consequences: more extensive ice melting at the poles, more sea-level rise, island nations submerged, the extinction of one-quarter or more of the worlds species, more violent hurricanes and massive famine in places such as Africa.

(Some scientists even said the IPCC reports underestimate the effects of a warming world.) The report goes on to say that now we must stabilize greenhouse gas emissions by 2015 and must mostly stop emitting carbon dioxide by the middle of the century.

Global warming is as real as gravity. Here's why. The sun does not keep the earth warm; it supplies the energy for the job. We have known since the early 1800s that particular gases in the atmosphere (now called greenhouse gases) trap and hold heat to maintain an average worldwide temperature of about 15 degrees Celsius (59 degrees Fahrenheit). The atmosphere acts as a blanket wrapped around the planet to keep it warm. This is pure physics, and there is no debate about it. If the earth did not have an atmosphere, the sun's energy would simply return back to space, and the average temperature would be about -18 degrees Celsius (0 degrees Fahrenheit). Earth would be a frozen wasteland where life as we know it could not exist. If there were no humans on earth, there would still be a normal, stable greenhouse effect taking place. However, with humans in the mix, there is an abnormal, unhealthy, unstable greenhouse effect, which in reality, is simply too much of a good thing. This is an enhanced greenhouse effect, commonly called global warming.

The name "greenhouse effect" came about because what happens in a greenhouse where plants are grown is similar to global warming. The sun's short-wave energy enters the earth's atmosphere, and some of it passes right through to the surface (some reflects back to space). This heats the earth's surface and reradiates as long wave energy, which is infrared, or simply heat. Atmospheric gases (and clouds) trap this heat and reradiate it, and here too some makes it back to space. The net effect is a habitable planet. This effect is not exactly the same as what happens in a greenhouse, where the glass roof traps the heat and holds in the interior atmosphere, but it is similar. This very complex process could be called the reradiation effect, but what's in a name?

When we change the chemistry of something, then that something changes. Take a glass of fresh water and add salt, and the water is changed. Taste, freezing point and electrical conductivity all change. If we add salt to salt water, it too will change. If we dramatically increase the amount of heat-trapping gases in the atmosphere, something must change, and something must happen, because we have changed the atmosphere. It always was absurd to claim nothing would happen, and it is still absurd to make that claim today. One can argue about exactly what will happen, and to what degree change will take place, but to argue that radically altering the atmosphere will do nothing is going against basic physics and chemistry, as well as common sense.

The average surface temperature on Venus is blazing hot at about 900 degrees F. Venus is closer to the sun than earth, but this is not the only reason why it is so hot. There is a raging greenhouse effect on Venus because its atmosphere is over 96 percent carbon dioxide (CO_2). The earth's atmosphere is less than 1 percent CO_2, at about .037 percent. Doubling CO_2 from its preindustrial level of about 280 parts per million (ppm)—again, as of 2008 it is up by about 38 percent—would represent a tiny fraction of the atmosphere but still could be very dangerous for life. When people say we can't affect something as vast as the atmosphere, they are very wrong. The small quantities of CO_2 relative to the size of the atmosphere make it very easy to make big percentage changes in this quantity of CO_2. If we were talking about nitrogen, which makes up about 78 percent of the atmosphere, it would be hard to cause serious change. But it is easy to affect CO_2 levels. A similar scenario holds true for the other greenhouse gases, which are also found in relatively small atmospheric quantities, yet are major players in life-support.

Substantial increases in all greenhouse gases have already been seen from preindustrial times to the present, and a doubling of CO_2 by 2050 and even a tripling sometime in the near future after that are possible, and out-of-control global warming that ends life as we know it is also possible. Positive feedback mechanisms—some of which are happening now, some of which are yet to be fully triggered—can throw us into a mode in which even if we stop emitting all anthropogenic greenhouse gases, we will still face serious problems.

The world's oceans are a gigantic sink for CO_2. Some of the CO_2 is taken up by phytoplankton, and some of it is absorbed by the water. The oceans have a limit on how much CO_2 they can take in, and today they hold about one-third to one-half of anthropogenic CO_2. This is no small favor, as global warming would be worse right now if not for the oceans' CO_2 absorbing power. However, mostly since about 1970 the world's oceans have been noticeably warming as they absorb heat from the warming atmosphere. As the water warms, its ability to hold CO_2 diminishes, and it may give that CO_2 back to the atmosphere in dangerously large quantities (the positive feedback would lead to accelerated global warming).

If rainfall increases into the oceans from global warming (as it has in some areas), adding to fresh water already put into the system by melting ice, surface waters become less salty and therefore less dense. This may decrease the downward movement and mixing to the deep ocean of CO_2 (and heat), further decreasing the ocean's ability for CO_2 uptake. The result: more positive feedback, and so more global warming.

Water vapor varies considerably in the atmosphere as it constantly moves into and out of the atmosphere, but it is the most powerful greenhouse gas. At the equator direct sunlight evaporates huge quantities of water into the atmosphere, causing a super-greenhouse effect—a natural process. However, as the oceans and atmosphere warm everywhere, more water will evaporate, and a warmer atmosphere can hold more water vapor, so more positive feedback for global warming is likely. Since the 1980s an increase in water vapor has been seen.

Gas hydrates (methane trapped in frozen water) exist in gigantic quantities in some deep ocean sediments. Warming oceans may release this methane into the water and then into the atmosphere, (meaning more positive feedback) and accelerate climate change to extreme and disastrous levels. That process may already be starting. There are also vast quantities of gas hydrates stored in the Arctic tundra permafrost that will be released if the tundra warms enough—and it is warming, so the process may be starting there too. Methane concentrations have not been higher than they are now in the past 420,000 years, and have over doubled from preindustrial times, and methane is a much more powerful greenhouse gas than CO_2.

Soils contain microorganisms that breakdown organic matter and release CO_2. Increased temperatures cause the process to speed up, making wetlands, farmlands and the tundra additional sources of greenhouse gas. About 15 percent of the global soil carbon is stored in wetlands alone, and some of this soil carbon will be sent to the atmosphere as CO_2 as well as methane (CH_4) due to higher temperatures. The Arctic holds vast quantities of organic matter frozen into the permafrost, but now the permafrost is melting. This is causing microorganisms to convert some of this matter to CO_2, potentially in massive quantities—and a similar process is also adding CH_4. It is estimated that hundreds of billions of tons of carbon is stored in the northern latitudes as organic matter. No one knows exactly how much of that matter will convert back to CO_2—carbon plus oxygen—as things warm up, but it may be dangerously large quantities, which will again be disastrous for life. Also, as the Arctic warms, entire forests are being damaged as their root systems literally start to lose ground. The term "drunken forest" has been used for trees that are starting to tilt. A beetle infestation brought on by warmer weather has killed or damaged over 25 million weakened spruce trees in Alaska alone. As forests and plant life start to die off, additional decay will take place and more CO_2 will be released. In general, if extreme events such as droughts, forest fires and storm intensity increase worldwide, then more trees and other plant life will die, and more CO_2 will be released. And even if trees are replanted or regrow naturally, they may not return fast enough to remove CO2 in time to stop runaway climate change. Some forests require decades to mature. Also, in higher temperatures, soil nitrogen will convert to atmospheric nitrous oxide (N_2O) another greenhouse gas. All of this creates more positive feedback.

Now, as the planet warms, it will warm even more because of what may be the ultimate positive feedback mechanism: the added warmth causes more and more snow and ice to melt. This is changing the reflectivity—called albedo, as I already mentioned—of the planet, which is normally about 30 percent. Again, that means normally about 30 percent of the sun's energy is reflected back to space—collectively from the surface, the atmosphere and clouds, and that means 70 percent of the sun's energy is absorbed. Part of this average albedo is due to shiny surfaces like ice and snow in the Arctic and Antarctic. This is one reason why these areas of the world are so cold. Glaciers, snow covered mountains and frozen water bodies also contribute to the process of reflecting energy away from the planet. Ice and snow have albedos of 80 to 90 percent, so as they melt, darker, much more absorbing surfaces such as water and soils are exposed, some with albedos as low as 5 to 10 percent. Albedo decreases and more energy is absorbed, and so more ice and snow melts, and so more energy is absorbed—and the process runs away with itself: more positive feedback. (This is a major reason why the Arctic is warming so fast.)

One of the most bizarre potential feedbacks is that if we stop burning fossil fuels, especially coal, then we actually may face a serious rise in temperature. The pollutant of concern is sulfur dioxide. When it enters the atmosphere, it reacts with other compounds to become a high-level sulfate haze. This haze is extremely reflective of incoming solar energy; therefore, this haze is keeping the planet cooler than it would otherwise be. Initially, it was thought that the cooling effect of sulfate haze was not that significant. New research shows that it may be very significant. The average residence time of the sulfates is only about a week; however, humans constantly produce it, so it is always in the atmosphere. If we stop generating this pollutant, it will quickly leave the atmosphere, and the true effects of global warming will be seen. Sulfates may be diminishing the effects of global warming by as much as 20 percent!

Global warming is complex; therefore there may be more positive feedback surprises. Collectively the ones I have described have the potential to throw this planet into an unstoppable global-warming event as more greenhouse gases are piled into the system. Global warming is taking place much faster than previously thought, as is seen in many events from sea-level rise to the rate of melting ice. We may only have 10 or 20 years left to change our ways before it is too late to avoid radical life-support changes, or put simply, the end of life as we know it. The time frame certainly may be longer, but what do we gain by taking the chance?

The anthropogenic greenhouse gases of concern are water vapor, carbon dioxide (CO_2), methane (CH_4), and nitrous oxide (N_2O). The source for water vapor has been covered. CO_2 comes from burning things, especially fossil fuels and their derivatives—oil, coal, natural gas, gasoline, etc. When burned anything with carbon in it will produce CO_2. Burn garbage or paper, or even light a cigarette, and out comes CO_2. Industrial sources and transportation produce the majority of CO_2 emissions. Biomass burning and deforestation through cutting and burning, as well as grassland burning, produce much of the balance. Forests are a warehouse of carbon storing as much as 100 tons per acre. Residence time in the atmosphere for CO_2 is about 120 years.

By some estimates, methane (CH_4) is accumulating in the atmosphere faster than CO_2 and it is over 20 times more powerful as a greenhouse gas. As already mentioned, atmospheric quantities of methane have more than doubled since preindustrial times. It is released from the stomachs of ruminant animals (livestock), from manure, wetland rice cultivation, solid waste landfills, oil and natural gas production and pipeline leaks. Atmospheric residence time is about 12 years.

Nitrous oxide (N_2O) originates from fossil fuel combustion, burning biomass and the breakdown of agricultural fertilizers. It is a more powerful greenhouse gas than methane and atmospheric residence time is about 115 years.

CO_2 produces about 60 percent of the global warming effect, while collectively the rest of the gases finish off the process. Over the past 10 years the growth rate for CO_2 has been faster than any 10 year period since the 1950s, when monitoring began. The majority of 20th century warming has taken place during the past 50 years. Average global temperatures could rise as much as five degrees Celsius (about 10 F) over the next 100 years as a consequence. As already mentioned, a five degree Celsius change in average temperature can be the difference between the beginning and the end of an ice age.

Heat is energy, and its movement around the planet via the oceans and atmosphere is the major reason for wind and ocean current circulation. The constant exchange of energy between oceans and atmosphere is a major factor for climate and life as we know it. Only the top 10 feet of the ocean's surface holds more heat energy than the entire atmosphere. (The entire ocean has a thermal capacity of approximately 1,600 atmospheres.) We are adding heat energy to both the oceans and atmosphere, which is affecting the movement and distribution of heat energy around the planet: therefore the earth's entire system is changing. This is a very dangerous thing to do.

There are other greenhouse gases that are rarely discussed. Chlorofluorocarbons (CFCs) and hydro-chlorofluorocarbons (HCFCs) used as coolants in air conditioners and refrigeration are also very powerful greenhouse gases. Compared to a CO_2 molecule, these gases can be many times more effective at warming the atmosphere. Certain hydrofluorocarbons, like HFC-134a, which is very commonly used, is over 1,000 times more powerful as a greenhouse gas than CO_2. Residence time in the atmosphere for these coolants is 50 to 100 years.

Low-level ozone produced by vehicle emissions as well as other sources, is also a greenhouse gas that has been increasing as more and more gasoline is burned. It is a potent greenhouse gas that is adding to global warming.

Briefly, here are some of the global warming effects—some of which I've already covered—not necessarily in order of importance.

Sea-level rise is being caused by melting snow and ice (glaciers, ice sheets, etc.) and by thermal expansion of the oceans as they heat up. Thermal expansion is responsible for much of the observed sea-level rise. In part because it absorbs more solar radiation, the Arctic is heating up much faster than the rest of the planet, including the Antarctic, and if the Greenland ice sheet completely melted, sea level would rise by about 23 feet. If both the north and south pole (Arctic and Antarctic) substantially melted, the sea level could rise 250 feet. I am not implying this will happen. However, a catastrophic sea-level rise of three feet in 20 years and about 15 feet in a century is not impossible. Evidence shows this may have happened some 14,000 years ago even without humans accelerating the process. Further, according to the Union of Concerned Scientists, some computer models show that "warming between two and seven degrees Fahrenheit above today's global average temperature would initiate irreversible melting of the Greenland ice sheet."

The Arctic is degrading more than the Antarctic, which as already mentioned is starting to show signs of stress and melting. Recent studies show that Arctic sea ice is melting faster than anyone had predicted. It had been thought that by 2100 summer sea ice would be gone. Then the ice started melting faster than predicted, and 2050 became the new estimate. Now the ice is melting so fast—to record lows—that some scientists are talking about the sea ice being gone by 2030! Then in December of 2007, after reviewing new data, a NASA climate scientist said the ice could be nearly gone by 2012. Time will tell.

Sea ice that is already displacing water does not directly contribute to sea-level rise when it melts, but its melting allows land ice to break free and melt into the oceans, and that does raise the sea level. Additionally, sea ice melting contributes to sea level rise because the dramatic change in albedo causes extreme positive feed back.

Sea-level rise could be three feet or more in this century. This would cause serious damage in coastal areas that are home to half the world's population. Even a six inch rise can lead to destruction of a structure that is well above sea level because coastal erosion and storm surge are enhanced. Many coastal areas and barrier islands will disappear, and coastal wetlands, bays and estuaries, vital to most marine life, will be dangerously altered both physically and chemically, as saltwater inundates them. This habitat destruction will also detrimentally affect other species such as turtles and sea birds. Additionally, the economic value of coastal areas is enormous and will be lost. Protection from hurricanes provided by barrier islands and wetlands (also coral reefs, more later) will be lost, exposing large populations to severe damage. These processes have already started. Even cities as much as 100 miles inland can be disastrously affected—including Washington D.C.—as water follows bays and rivers inland. Many small island nations around the world will vanish or be seriously altered. Even if greenhouse-gas levels are stabilized, sea level will continue to rise, possibly for more than 100 years, as the oceans have a slow response time.

The world temperature will also continue to rise before stabilizing, and remember, disease carrying insects thrive in a warmer climate. Mosquitoes, flies and even water snails will thrive and expand their range, and diseases such as malaria, dengue fever, yellow fever, schistosomiasis, lymphatic filariasis and others may increase. Increases in asthma and chronic lung disease due to increased low-level ozone are possible. More allergies and dangerous algal blooms are also possible.

Worldwide, climate and weather are being affected. Seasons are changing; in the Northern Hemisphere spring starts earlier and fall starts later. Heat waves, droughts, floods and forest fires are taking place with frightening frequency and intensity, and many records have been broken (heat records continue to be broken through 2007, and from 1993 to 2003 in the U.S., in six out of 10 years, heat was the leading weather-related killer). Precipitation patterns are also changing: erratic weather may become the norm. More frequent El Nino events have taken place over the past 15 years. (El Nino causes a temporary change in ocean and atmospheric circulations in the tropical East Pacific that catastrophically affects weather in many areas around the world.) In some areas more salinization and desertification of agricultural land will take place. Salinization is a process in which the soil becomes saltier because of evaporation. Desertification can be severe when fertile land becomes desertlike or when a desert expands. Both processes reduce or end plant growth.

More intense storm activity such as hurricanes is likely. (Hurricanes, typhoons and cyclones are all similar, the difference being where they form: the Atlantic and northeastern Pacific, the northwestern Pacific, or the Indian Ocean respectively.) There is still some debate, but more evidence is showing that hurricanes may be much stronger in the future. Hurricanes build by drawing the heat energy and moisture out of the oceans. During 2005 sea-surface temperatures were at record highs as global warming heated the oceans. As already mentioned, in 2005 hurricanes broke records, and the increase in North Atlantic hurricane intensities over the past 30 years correlates with increasing sea-surface temperatures. However, during 2006 the number of storms in the North Atlantic came in below predictions. A recent study shows this lull was due to El Nino and La Nina (a reversal of El Nino) events that changed wind patterns and cooled some areas of the ocean, changing tropical Atlantic conditions for hurricane formation. It is also possible that fewer hurricanes will take place because the ones that do form will be much larger and more powerful. This may be because the larger storms draw so much heat out of the oceans and leave a cooler ocean behind. We will have to wait and see what happens as we test the world's tolerance with global warming.

About 60 percent of the U.S. has experienced drought since the late

1990s. In October 2007, northern Georgia declared a state of emergency due to decreasing water supplies. Further, 36 states may face water shortages within five to 10 years. Water shortages due to changes in the hydrologic (water) cycle causing less recharge to aquifers and reservoirs, will become more severe. Changes in precipitation—rain and snow—are beginning. In mountainous areas in the western U.S. and Alaska, and in many other areas of the world, snowpack melt is needed for drinking water (also for electricity from hydroelectric sources); therefore changes in snowpack will negatively affect drinking water supplies for a large portion of the world's population. Drinking water problems caused by sea-level rise inundating above-ground water supplies is possible. Salt water intrusion can also damage underground water supplies (aquifers) that are near coastlines.

In a worst-case global-warming scenario, a collapse of worldwide food supplies is possible, although initially we may see an uneven change in agricultural yields. Some areas will get more food, some less.

Mass species extinction—terrestrial and marine plants and animals—continues and climate change can spin this out of control to the detriment of ecosystems. Die-off of coral reefs—home to 25 percent of ocean marine life—caused by sea-level rise and warmer oceans has started. Add to this human activities such as coastal development, pollution, overfishing and destructive fishing, such as using cyanide to catch fish, and the problem gets worse. Coral reefs are in severe decline—as much as 25 percent are already ruined, and we may lose them all in this century. One billion people in Asia alone rely on the fish that coral reefs support. And increasing ocean acidity (a process called acidification) may very negatively affect the oceans as anthropogenic CO_2 combines with the water to form carbonic acid. This is detrimental to marine life in general, especially coral reefs, and dramatic changes to the marine environment due to increasing acidity may already be underway. Surface water acidity has risen and may continue to rise to deadly levels by the end of the century or sooner. Further, reefs act as coastal barriers to storms. This vital protection may be lost. Ocean acidification may eventually seriously effect phytoplankton (and zooplankton), the base of the marine food chain and a major source of atmospheric oxygen. Additionally, heating the oceans and changing the quantities and patterns of fresh water coming into the oceans from rivers and runoff, also may be negatively changing the distribution of marine species. As I already mentioned, we may lose the oceans as a food supply over the next 40 years!

The slowing or stopping of the ocean conveyor system, which could radically change climate, is also possible. Changes in the ocean conveyor system were portrayed in the movie *The Day after Tomorrow*. The movie exaggerated what would happen if the system were affected, but the premise was based on actual research. Also called the thermohaline (temperature and salinity) circulation, the conveyor system is a massive, worldwide movement of water and heat from the surface to the deep oceans and back to the surface again. It is a critical part of the climate system.

At the equator the sun's energy heats the ocean's surface. This warm water moves northward up into the North Atlantic, releasing its heat to the atmosphere. Evaporation takes place, and moisture is also released to the atmosphere. This cools the surface water and makes it more salty; therefore the water becomes more dense and sinks in gigantic quantities. One estimate is that the amount of water moved in this process is more than 20 times that of all the world's rivers combined. The water sinks to depths of 4,000 meters (13,000 feet) in the North Atlantic and then flows southward along the bottom, then to the east around the southern tip of Africa. There it hits Antarctic waters and flows into the Indian and Pacific oceans, still along the bottom, eventually rising up to again join the surface circulation of the Indian and Pacific ocean areas, where it picks up more heat. It then heads back the way it came for another trip across the equator and up into the North Atlantic. The round trip may take as long as 1,000 years.

This system acts as a thermostat for the entire planet, keeping winters in the Northern Hemisphere relatively warmer, and equatorial areas relatively cooler. It also delivers important moisture to the atmosphere for rain and food growth in areas of the Northern Hemisphere. If enough ice melts into the North Atlantic, then fresh water starts flowing across the ocean's surface. Additional rain from global warming may also add to this. The fresh water reduces the density of the cold, salty water, and the conveyor system slows or stops because it is density driven. Satellite studies and other work have shown that, indeed, as Arctic ice and glaciers melt, salinity in the northern ocean has declined, and currents have weakened. If the ocean conveyor were to stop, a massive distribution of heat energy would stop. The Northern Hemisphere should then get much colder while the equatorial regions heat up. It is believed by some that this could throw the world into a cooling trend, possibly an ice age, and one study suggests the conveyor has already slowed possibly by 30 percent. Evidence suggests that over the past 75,000 years the conveyor system has been changed more than once by fresh water entering the North Atlantic, which caused very abrupt climate change. The worst case is a doomsday scenario that would shatter human existence as we know it—as well as that of other animals—as agriculture and other plant life collapse and livable climate becomes unlivable. Interfering with ocean circulation could cause major climate change over a very short time, possibly only 10 to 20 years.

This theory is new compared to global warming, and more research needs to be done before any conclusion is reached. Yet, there is no question that if the circulation changes or stops, something will happen. It's just plain old physics, as heat flow is radically altered. However, exactly what would happen is in debate. Can we put enough fresh water into the North Atlantic to stop the circulation? Will a warmer atmosphere due to global warming cancel out some of the effects? There are many unanswered questions.

If there is the remotest possibility that human-induced changes can stop or alter this tremendous movement of heat around the planet, then this is another reason for us to change our ways, because the unknown is not always a fun place to go.

11
THE END

Starvation, disease and war are rampant around the world. The economically strong are surviving well, and the economically weak not so well. How can we consider human fitness a success and better than the Darwinian world if it continues to fail miserably for at least half the world's population, and if human fitness continues to be self-destructive by doing so much damage to the very things that keep us alive?

If we are to become truly civilized and climb to the next level of intelligence and protect the world we live on, then we must change. For humans to truly be humane, to no longer have wars and maintain weapons capable of destroying most life, and to no longer be inhumane to animals—including the human ones—is probably our next step up, and it's long overdue. Humans are on a journey to this higher level of development, and whether we get there will be determined by how we deal with our current problems. We have not yet gotten to the point at which we use our intelligence more for humanity and survival, than for our greed, wants and desires.

We are a very new species compared to so many others. We are far from perfect—though it may be fairer to say we are far from complete—and so we have a ways to go to that realm where we protect all of our population as well as the entire planet and maintain complete respect and peace for all species. It would be a shame if along the way to recovery we destroyed ourselves, because we do have the potential to get to this higher plateau.

People around the world have become much more conscious of environmental damage. Even the oil and car companies now admit that global warming is real, and some of them have run newspaper ads and television commercials stating how environmentally aware the companies are and how they are working on the problems. We keep hearing the term "green" as corporations and politicians start to change their acts. But without a true understanding of what care for the environment entails, the term "green" is in many cases just a good advertisement to make more money. For sure, some companies are greener than others, but the term has been misused and abused.

An April 2007 *New York Times* article makes the point. The headline reads, "Bloomberg Draws a 25-Year Blueprint for a Greener City." The money needed will be hundreds of millions of dollars and the fight is on with the state and federal government to get it. The mayor wants to charge fees for driving a car into Manhattan, improve mass transit, improve efficiency in power plants and buildings, create land for housing, add bike paths, cultivate mussels to reduce river pollution and eliminate sales tax on energy-efficient vehicles. This is all great and commendable, except for the main reason it's all being done. The article starts out by saying, "The plan is intended to foster steady population growth, with the city expected to gain about one million residents by 2030."

The American public (other societies too) and the corporate and political leaders are coming out of the dark too slowly and have only scratched the surface of fixing the problems with ideas such as population control, new technologies, conservation, respect for other species and education. And the coming out is only because problems like global warming have been thrust on us because we ignored the issue for so long.

Many people consciously or subconsciously assume that the earth is in place solely for the demands of humans, while not comprehending that those demands are completely unreasonable. Foresight has not been the order of the day; instead we live in the moment of now, and now we have a problem, and now we have to fix it. However, we haven't yet been grabbed by the understanding that at the moment we decide to fix a problem, that the problem may be beyond the point of no return.

State and local governments—which are way ahead of the federal government—are making changes such as using hybrids as county vehicles, improving energy efficiency and building codes, pushing renewables and decreasing pollution. Ten states, along with environmental groups, have even sued the EPA over global warming. The courts upheld a ruling that the EPA must regulate pollution that is harmful and that CO_2 is a harmful pollutant. Specifically, Massachusetts won its case when the court agreed that Massachusetts's coastline was already damaged by climate change and that further damage could be expected. This was a big step and long overdue.

On a federal level our representatives are starting to attempt change. The Climate Security Act of 2007, was approved by the Senate Environment and Public Works Committee. This bill (S.2191) is extensive and has been called by several senators the Manhattan Project for energy (but it is definitely not), and indeed it is a far reaching bill to reduce greenhouse gas emissions—while at the same time specifically allowing for continued robust growth of the U.S. economy. Large sums of money and resources would be allocated over many years for renewables such as solar and wind, but greater amounts of money and resources would go towards allowing continued burning of fossil fuels—in a cleaner manner—and, to use carbon sequestration in the coal industry. It also advocates increasing the use of nuclear power. Changes to the bill are being demanded by politicians and industry, and the bill faces major hurdles to become something that can make a difference, before becoming law—if it does. In 2008 the Act was still being worked on.

In any event, time is not on our side, and there is still a serious lack of understanding about the connection between humans and the environment. Eventually, maybe the American people will be the most environmentally conscious people on earth and save the planet with our corporate and government machine. This is possible; we do have the technology. But will we do it?

Our system has lost its sight, so how can we pass vision on to our children? Many first-year college students that walk into my classroom have a problem with basic math and English and have little concept of science. Many studies on the math and science ability of U.S. students show them falling behind much of the world. In 2004 the *L.A. Times* ran a piece titled "With Science, U.S. Is Getting Left in the Dust." The American Federation of Teachers posted an article in 2005 entitled "Students Rarely Ready for College." A study funded by the Pew Charitable Trusts and reported on by CNN in January 2006 was headlined "More than half of students at four-year colleges, and at least 75 percent at two-year colleges, lack the literacy to handle complex, real-life tasks such as understanding credit card offers." And the February 2006 cover of *Time* magazine portrayed a comical picture of a teenage student having blown himself up in a chemistry lab. Guess what that article was about?

Drugs have been on the street since I was a kid, although our government still continues to claim they are trying to stop this. Two working parents, materialism, negative music, gratuitous and excessive sex and violence on television and in films and video games, don't help either. Internet chat rooms, constant e-mail, and porn sites don't help either. And cell phones and music devices constantly glued to our kids' heads are destructive to their focus and concentration. Computers have become a distraction, doing everything for them and diminishing their ability to read, concentrate and study. Too many students refuse to read and won't take notes. They prefer typing the question into a search engine and printing out the answer. For a price they can download a paper or report on any topic, print it, and hand it in, and many of them do this.

Students want to consume, and materialism and consumption are so ingrained into their spirit that imagination, creativity, curiosity and critical thinking have taken a back seat. These factors and more have contributed to our children's demise and therefore ultimately to our society's demise. These are the types of things that stunt our thought processes and allow a society to become ignorant of, or indifferent to, science and environmental issues. Educating anyone about the environment is not a priority, and if we don't even do an adequate job in basic education for our children, then how can we enlighten them, their parents, and the public about the environment?

As already mentioned, there are ways to start the process, such as regular public-service television announcements and government officials, like a president in a State of the Union address, telling it like it is, but this type of enlightenment is not yet happening. By not teaching and enlightening, we have been sending a message to our society that it's okay to take everything that keeps us alive for granted. Air, water, soil, plants, animals and climate are permanent fixtures for human beings.

We wallow in our own cultural, political, and religious beliefs as though this is all that matters, and we think that these beliefs will solve all our problems. Instead, as the most intelligent species, we need to face our responsibility to the other life forms and to take care of them and everything on the planet down to the last bacteria. And if we do this, then automatically we will be taking care of ourselves.

In order for a leader to lead, he or she must be realistically educated about the environment—and must be honest. However, neither Democrats or Republicans tell the whole truth. Elections have become a farce in which lies and half-truths dominate, and in which little is adequately explained, while great promises are made and too often not kept. We are not truly being represented, and the problem affects the environment.

One ex-politician raising serious concern about global warming and educating the world on the issue is Al Gore. What he has said in film and in print has been stated (in more detail) countless times before in college classrooms, research facilities, documentaries and books. The difference is that celebrity gets the message out in a broader, more powerful way. I hope he understands all the other ramifications of our culture's population and economic system.

The pace of day-to-day living has become rapid, requiring cell phones and pagers and computers so that we can do more, produce more, work more and sell more in any given hour or day. We are not only stressing ourselves out and decreasing our quality of life and in some ways our health, but we are reducing life-support on the planet. Is our purpose to dominate all other life forms on the planet and expand until they are mostly gone, while somehow we remain viable? If human beings don't have the same purpose that other species have—that is, to keep themselves and all the ecosystems healthy and functioning—then we have no purpose other than a self-centered one that is self-destructive.

From the time that we are old enough to see and hear, we are marketed to and manipulated to such a degree that even as adults many of us don't really think for ourselves. Given all the knowledge that we have on global warming and how it threatens our existence, I was astounded when I read two articles, one in the *New York Times* and one in a local paper.

The first article, from the *New York Times* in October 2005, was entitled "As Polar Ice Turns to Water, Dreams of Treasure Abound." The article acknowledged that global warming is significantly melting Arctic ice in this northern section of the world. Glaciers are melting, ice sheets are disappearing, and mammals are threatened. However, the article says, there is a positive side to global warming. New shipping lanes will be opened because of the melting ice, and areas once difficult or impossible to get to for fossil fuels now can be accessed and the fossil fuels extracted. The melting has created a bonanza of opportunities for us to get more oil and gas from virgin territory, and the prize will be hundreds of billions of dollars to the governments and corporations worldwide that are currently fighting over territory rights. The article went on to say that now fishing areas can be extended further north for more money to be made. The Arctic is being cut up.

What about the fact that the oceans are in serious trouble from the already depleted fish stocks? What about the polar bears, foxes, seals, walrus, birds and other species that are at extreme risk of serious decline? Is the chance for more money to be made and more fossil fuels to be burned all that global warming means to our economic world, so that more global warming will melt more ice? A true understanding of the Arctic melting would be sending such shockwaves of fear through the economic world that people would be climbing over each other to get to solar and wind power—and shipping ice cubes to the Arctic by the boatload. The Arctic melting potentially represents catastrophic events for humankind, the beginning of a runaway climate change that very soon will make full human existence impossible. Yet the article represented this climate change as the new California gold rush. In 2007, for some, it was still viewed this way.

In the second article, when citizens were asked if we should drive SUVs, one gentlemen protested vigorously that he had the right to drive anything he wanted because he could afford the gas and liked a big vehicle, and wanted his kids to be protected. Nothing was said about the people—children included—who are killed or injured specifically because of SUVs, nothing about global warming, acid rain, air pollution or the associated effects of burning so much gasoline.

These articles were written at the end of 2005, when global warming already was very well understood. Are these articles a statement about the overall condition of our society's psyche? Or are they just representative of a smaller group? Since the articles were written, have we changed enough? In 2009 and 2010, and further into the future, will these types of ideas continue? Ignorance is not bliss; it's very dangerous.

I'm amazed when I hear people say how great a housing boom is, like the one that started in the late 1990s. When the CEOs of the major home builders talk, they talk of their ability to grow their business and build new houses with no end in sight. They have clearly found a world of limitless space, and a world that can absorb unlimited garbage, pollution and ecosystem damage.

We will keep building more of what we need to accommodate our growing population. Highways are widened all over the country, and when traffic jams up again, they will be widened again. This is the way the American culture operates—and so does much of the world. Wouldn't it be a shock if someone in power mentioned reducing population and consumption as a solution?

Even as environmental issues become mainstream—as the September 2004 issue of *National Geographic* and the April 2006 issue of *Time* magazine illustrate, along with so many television programs—things are changing much too slowly. A radio survey in August 2004 was enlightening. A number of people were asked what they felt was the greatest threat to our economy. The number one response was terrorism. In 2006 similar polls showed that the public was concerned about terrorism, jobs, education and retirement, with little mention of the environment. In 2007 the environment got a bit more play. In 2008 economic issues pushed environmental issues far into the background. Ultimately, the number one threat to all of us and to our economy is degradation of the environment.

I may be too idealistic, and I am well aware that the world will not change overnight, but just imagine for a moment that the vast majority of the world's population fully understands and appreciates the environment? Could we then end wars and cultural and religious differences through environmentalism? Ultimately, the conflicts that exist from culture to culture stem from ignorance. Again, the earth does not care about our religious or cultural beliefs; the earth will roll right over all of us and our beliefs. If we all had common ground, so that we were in fear for the planet and understood that our actions were threatening our very existence, then we would be less likely to do more damage. We would realize that the effects of an exploding bomb on the planet's life-support systems are the same as the effect the bomb has on the people who get maimed and killed every year by wars large and small. The insanity and barbaric happenings might slow down, maybe even stop. The world's people are in reality all one tribe, and if we don't start acting as one, the trouble we face will get worse.

We must understand that the earth does not owe us a living. The earth owes us nothing, not air, water, food, cover or space. We must shed our arrogance and get humble in the face of the other animals that are showing themselves to be smarter than us. It may be that we don't respect ourselves because we don't respect them. Until we recognize this concept, we have a problem.

Many cultures throughout history have seriously declined due at least in part to environmental problems related to overpopulation and overuse of resources. Easter Island and the Maya civilizations are two examples, yet they were tiny populations and remote cultures compared to the world today. They grew and expanded as far as the technology of their day would allow them, without the full understanding of what the expansion of their population would do to their food, water, soils and forests.

Today's modern technology has allowed humans to expand our presence over the entire planet, but no amount of additional technology will allow us to violate the laws of physics and nature. The difference today when compared to past cultures is that we threaten the entire planet, and thus we threaten a complete disintegration of the world's ecosystems and humankind, whereas the smaller cultures only threatened a fraction of the planet and its future. The other major difference between past and present is that we have the technology and knowledge to preserve the planet and all of its inhabitants. Will we use these abilities, and if we do, will it be in time?

I believe humans have a fifty-fifty chance of surviving in a modern, civilized manner. And I am frightened because we are arguing the probabilities of severe problems and/or human demise, since we have casually accepted the possibility of these things because of what we have done to earth. At the same time, we attempt to minimize the problems so that we can continue doing what we desire, what we consider normal. We are on the edge, and we are seriously pressing our luck. Nothing is written or ordained that says humans must win the game. Earth is like one grain of sand on all the Hawaiian beaches. This one grain of sand removed, would affect those beaches about as much as earth's removal would affect the cosmos.

Thousands of leases have been issued to drill for more oil and gas in the northern U.S. on pristine and untouched areas of Montana and South Dakota and elsewhere. These are the last sections of this great country that haven't been ravaged by overpopulation, but deterioration is starting as roads cut in, habitat is fragmented and pollution and building begin. The struggling yet functioning ecosystem of antelope, elk, wolves, big horn sheep, grizzly bears, lynx, wolverines, mountain lions and others will suffer. The sad fact is that the amount of oil and gas in these areas isn't all that substantial, but the extra resources will allow us to expand our economy a bit further. If we realistically conserved resources, the resources in these areas would not be needed, but so far we simply will not seriously conserve until we are forced to conserve—or until energy gets too expensive.

A good reason to conserve would be to buy time while we switch over to solar power and lower our population. Instead, we worry about our children's health while we give them lung cancer and asthma attacks because of what we drive and the way that we live. Instead, we plan for and worry about their future while we simultaneously take that future away from them. Humans have become a positive feedback mechanism for our own demise.

So, I can tell you for sure that we have not been operating as the most intelligent species, and Albert Einstein and Charles Darwin were not idiots. The laws of physics and nature rule, not the laws of economics and human culture. If we do not stop bickering and fighting over our religious, cultural and political differences, we will not survive. If we continue to crave material wealth and items far above what we need to live, and if we continue to let our children's minds be molded by what has become a commercial, synthetic world, then we are in trouble. Someone once said, "We live in a world of lies. The Politicians lie. The business leaders lie. Madison Avenue lies. If you buy the perfume you will not get the date."

Someone else once said that the meek shall inherit the earth. It may be that the meek will be the other life forms. If we are not very, very careful, then we are going to soon change this planet so dramatically that it will not support us in our current form, and our population will crash. Our future really can turn out like a scary science fiction movie in which there are only small pockets of us and them left in a decimated world. I wonder how "they" will feel about "us" if Darwinian fitness once again fully dominates.

PART TWO

12
OZONE SHIELD BREAKDOWN

Our atmosphere is made up of layers. From the surface up to about 11 km (6.8 miles) is the troposphere, then there is a boundary or pause called the tropopause. From there the stratosphere extends up to about 50 km (31 miles), where there is another pause (stratopause) and then more layers, the mesosphere and thermosphere extending much further up. Temperature decreases as altitude increases in the troposphere, while in the stratosphere as altitude increases temperature increases (but it's still very cold). This temperature increase is due to ozone absorbing the sun's energy, and the warmer air on top of colder more dense air causes the stratosphere to mix and move horizontally, which is very important to the properties of the troposphere. It is in the vertically mixing troposphere that our climate and weather get sorted out, and in the stratosphere that the importance of ozone comes into play.

Ozone (O_3) is most concentrated at about 25 km (15.5 miles) above earth's surface, and is spread out above and below this. Ozone is a form of oxygen, and in the stratosphere it is vital to our survival. In the troposphere (there is some naturally occurring ozone) ozone is produced from burning fuels such as gasoline, and it is not only a greenhouse gas but also a serious pollutant detrimental to human health (more later). Ozone is ozone, but tropospheric ozone and stratospheric ozone development are two different processes not to be confused.

Ozone in the stratosphere is constantly formed and broken down by the sun's energy. When an oxygen molecule (O_2) is hit by the sun's energy, it is split up (through photodissociation) to form two oxygen atoms (O). Then another O_2 molecule finds the single oxygen atom and hooks up with it to form ozone (O_3). Two plus one equals three. But the sun's energy eventually breaks up the ozone, and then the process starts again. This process maintains a stable amount—a constant quantity—of stratospheric ozone, which keeps life on the planet alive and well. There is much more to the complex formation and breakdown of ozone, both natural and unnatural, but these explanations will suffice.

The quantity of O_3 in the stratosphere is about 12 ppm (parts per million), which is about .0012 percent. This is substantially less than CO_2 levels of .037 percent and again underlines how easily we can affect these gases, adding or subtracting from them, because their quantities are so small to begin with. Ozone is every bit as powerful and important as CO_2. If we pushed all of it down to the surface of the Earth, where the air pressure is much greater than it is in the stratosphere, the thickness of the O_3 layer would be about 3 mm (millimeters), which is a little more than one tenth of an inch. Its quantity is nothing compared to the rest of the atmospheric gases, yet it absorbs 99 percent of the ultraviolet radiation (UV) that is deadly to life. The three different types of UV are A, B and C, and they vary in damaging intensity from A, the least, to C, the most. Here I will simply use the term UV.

If the ozone shield, as it is rightfully called, were completely destroyed, then life on the planet for the most part would end as all the incoming UV radiation made it to the surface. UV is very damaging to animal and plant tissues, therefore most plants and animals, including humans, cannot live exposed to high levels of it. Eye cataracts, skin cancer, and weakened immunity are the results of excess UV exposure in humans. Protein and DNA molecules on the surface of living things are damaged by UV. Damage to terrestrial plant life—which again, is the basis of food, about half of atmospheric oxygen, and part of the climate system—has obvious consequences. All the interactions among all species would be very negatively affected.

In the oceans water protects most marine life from UV, but damage to fish populations that swim near the surface in the Antarctic has been seen. Damage to Antarctic phytoplankton—which live at the surface—has also been seen from UV exposure, and photosynthesis in phytoplankton is inhibited by UV. Again, phytoplankton is the base of the food web in the oceans, and we get about 50 percent of our atmospheric oxygen from them. UV may also be damaging coral reefs that are near the ocean's surface. Excessive UV exposure could easily collapse marine ecosystems.

Catastrophic ozone-shield breakdown would also change climate because ozone keeps a stable relationship between the stratosphere and the troposphere. If O_3 were gone, then it is possible that the two atmospheric layers would mix and turn into one gigantic vertically mixed layer, setting up dramatic changes for climate. Ozone breakdown might make global warming look like fun.

As hard as it is to believe, humans have seriously affected the ozone layer, but thanks to research, and specifically the work of two scientists, Sherwood Rowland and Mario Molina, the problem was caught in time, although O_3 breakdown is still a threat. In 1995, Rowland and Molina received the Nobel prize for their contribution to discovering the O_3 problem.

Ozone depletion over the Antarctic (south pole) was happening since about the 1960s, but ironically it was not recognized until later on in part because the amount of O_3 loss was so great. Instruments had been programmed to ignore large O_3 losses because this was thought to be impossible. But in 1985 a British research group stated that O_3 levels over the Antarctic were dropping significantly every year when the sun came back in September and October. This was the first report of an ozone hole over the Antarctic.

The hole is actually a severe thinning of the ozone layer by as much as 70 percent. In 2000 the size of the hole reached a record 29.2 million square kilometers (11.3 million square miles), an area larger than North America. In 2003 the hole was the second largest on record at 10.9 million square miles. In 2007 the hole was about 9.5 million square miles. Ozone levels have decreased worldwide, albeit to a much lesser degree than over the Antarctic.

The reason for the problem is human-made chemicals released into the atmosphere. In the late 1800s and early 1900s, refrigeration was primitive, when chemicals such as ammonia and methyl chloride were used in the systems, and fatalities from leaks took place. Around 1928 chlorofluorocarbons (CFCs) were invented, and the world of modern refrigeration took off. Over the years a variety of these very stable and safe chemicals (such as Freon) were introduced not only for use in refrigerators and air conditioners but also as foam-blowing agents for insulation and packaging, in the electronics industry to clean parts, and as aerosol-can propellants. Yes, hairspray and underarm deodorants in spray cans—eventually banned in the U.S.—contributed significantly to the problem.

Over five billion kilograms (11 billion pounds) of CFCs have been released into the atmosphere, and some of the effects are yet to be seen as CFCs are slow to breakdown. Broken off by the sun's energy (photodissociation again), it is the chlorine atom within the CFC molecule that destroys O_3. The chlorine is catalytic, meaning it promotes the reaction that destroys O_3 but does not easily get used up itself. One chlorine atom in the stratosphere can last 100 years and can breakdown up to 100,000 molecules of O_3.

The reason for the extensive and intense breakdown of O_3 over the Antarctic is because during the dark Antarctic winter, from June to August, the polar vortex (a whirling in the stratosphere) concentrates large quantities of CFCs and other gases in the Antarctic stratosphere. Cold temperatures allow polar stratospheric clouds to develop, and complex reactions on the cloud particles allow chlorine to build up. In September when spring arrives, the sun breaks up the clouds, and the chlorine is released in vast quantities, in essence shocking the system and destroying massive quantities of O_3, creating the famous ozone hole. In the Southern Hemisphere, not only the Antarctic but Australia, New Zealand and Chile are affected by the seasonal ozone hole, and citizens in those nations receive UV alerts, especially during the spring.

The most obvious effects of O_3 breakdown are unfortunate. One out of two Australians may develop skin cancer, and some of these cases will be fatal. In the United States the EPA estimates that 12 million Americans will eventually develop skin cancer caused by the loss of stratospheric ozone, and that over 90,000 of these cancers will be fatal. Approximately one million eye cataract operations take place every year in the U.S., and no one knows how many are due to excess UV exposure.

Other chemicals that damage O_3 are halons, methyl bromide, methyl chloroform, carbon tetrachloride and nitrous oxide. Halons are used as fire suppressants; methyl bromide is a pesticide; methyl chloroform an industrial solvent, along with carbon tetrachloride, which is also used in other industrial applications. Nitrous oxide, a global warming gas, comes from fossil-fuel combustion, burning biomass, and agricultural fertilizer breakdown.

The majority of the world's population lives in the Northern Hemisphere, and a large ozone hole there like the one that develops over the south pole would be a disaster. Because of differences between the Arctic and the Antarctic, it is thought that an ozone hole over the Arctic cannot happen. However, appreciable O_3 breakdown—as much as 25 percent—over the Arctic during cold winters has been identified, although it is not as severe as in the Antarctic.

Worldwide, ozone loss may peak by about 2010, and full recovery of the ozone shield is expected in about 40 years; however, there may also be surprises here as we continue to change the atmosphere. For example, global warming is trapping energy in the troposphere that isn't getting back to the stratosphere. This may be causing cooling in the stratosphere and adding to O_3 breakdown, especially at the poles. Global warming may actually delay the recovery of the ozone layer.

In Montreal, Canada, a meeting was held in 1987, and the Montreal Protocol was negotiated. This was an international meeting, and 184 countries eventually signed the agreement. Rules and regulations were laid out, and targets for the phaseout of CFCs were set in motion. During the 1990s amendments were added, as the seriousness of the situation grew. But in 1995 during the 104th Congress, three bills were introduced to put an end to American participation in protocols banning CFCs. In 2003, the Bush administration attempted to exempt the methyl bromide phaseout. Fortunately, both initiatives failed.

The replacement for CFCs, HCFCs, also damage the ozone layer, although to a lesser degree, and they too will be phased out, by 2020 in developed nations and by 2030 in undeveloped nations, according to a 2007 agreement at the United Nations Environment Programme, set up to toughen the Montreal Protocol. HFCs, another replacement, have no chlorine and do not damage ozone but are a powerful greenhouse gas, as are HCFCs.

CFCs, methyl chloroform and carbon tetrachloride have been phased out in the U.S. and some other developed countries. However, halon and CFC-12 still pose a threat to the ozone layer. Leaks from old vehicle air conditioners and old refrigerators continue, and Korea and China still produce halon-1211, while India, Mexico, and China have increased production of CFC-12.

Jet aircraft fly very close to and sometimes in the stratosphere and they contribute to ozone depletion, as well as global warming, acid rain, and air pollution. As world economies expand, we will further increase—possibly double again—the number of jet aircraft in the sky.

13
AIR POLLUTION

There is natural pollution from sources such as lightning-induced forest fires and volcanic eruptions, but the pollution of most immediate concern to the earth is human induced. This air pollution is made of gases, solids and liquids. When we burn things, manufacture things, create dust and particulate matter, we pollute the air. Anything that evaporates can generate air pollution, even house paint or perfume. The earth has the capacity to dilute some pollution that humans create, but not in the large quantities that we create.

Industrial smog has been typically called a mixture of smoke and fog. There are different types of smog, and commonly the ingredients are sulfur compounds, water vapor and soot. Air pollution can and does kill people. In October 1948, in Donora, Pennsylvania, an industrially induced pollution event took place that killed about 70 people and sickened 6,000. A thermal inversion (warm air on top of cold air) over a five day period caused a deadly buildup and trapping of industrial pollutants from a nearby steel mill that used high-sulfur coke.

Four thousand people in London died in December 1952 from industrial smog. Over the next two months 8,000 more died, possibly from the same event, though it is not known for sure. These were acute events. But the chronic exposure to air pollution that we are all subject to is of most concern. Breathing can be hazardous to your health.

There have been other pollution events over the years that have killed or sickened people, and as a result much stricter pollution laws have been instituted. In the U.S. the Clean Air Act of 1970 started the process of reducing air pollution, and over the years amendments added to it have been very beneficial. We are far better today than years ago at controlling air pollution. Cars burn substantially cleaner than just 30 years ago, and industries have been regulated. However, the problem of air pollution is still serious because as our population, economy and consumption grow, we use more manufactured items, and we burn more fossil fuels; therefore some of the positive effects of cleaning up what we do are negated. There is a point of diminishing returns for any environmental improvement if we keep adding to the problem.

While developed nations like the U.S. have made improvements, (though they still have air-pollution problems), in many developing nations air quality problems are severe. At times dust, soot and smoke in Chinese cities is ten times the level considered safe for human health. Beijing, China, has been rated the most polluted city of all the major metropolitan areas of the world, and pollution in general in China (air and water) is severe and sickens or kills people. Mexico City where during the winter months as many as 40 percent of the young children may suffer from chronic respiratory sickness—takes second place. Yet Los Angeles, California, generates some of the world's worst smog and has been designated the nation's most polluted city. The state has over 10 million cars and trucks, a large population, and a very large economy and industrial base all of which keep growing, and therefore growing pollution problems all over the state, not just in Los Angeles.

There are hundreds of known air pollutants, such as chlorine, hydrochloric acid, formaldehyde, benzene, vinyl chloride, asbestos and radioactive substances. They exist at varying concentrations. About 188 pollutants are known as air toxics, or hazardous air pollutants, and are emitted from different forms of transportation, from heating and chemical use in the home, and from manufacturing processes in small and large businesses where an array of fossil fuels and chemicals are used.

Transportation and industry account for specific sources of pollution, called primary and secondary pollutants. However, living in and running a home also contributes to this pollution. Major primary pollutants are particulate matter (PM), volatile organic compounds (VOC), carbon monoxide (CO), nitrogen oxides (NOx), and sulfur oxides (SOx).

Some of what makes up PM is smoke, soot, sulfuric acid, arsenic, asbestos, PCBs, oil, pesticides and carbon. These tiny particles, many of them microscopic, are suspended in the air, and many other polltants can attach to them. When breathed in, they find their way deep into the lungs and can be very harmful, causing respiratory and heart problems, and they can be carcinogenic (cancer causing).

Some sources of VOCs include incomplete combustion of fossil fuels, industrial and manufacturing processes, and evaporation of solvents and gasoline. They can also be hydrocarbons, and thousands of VOCs are known. Many are carcinogenic, and some are major contributors to the formation of low-level ozone.

Carbon monoxide (CO), a very abundant pollutant, comes mostly from the incomplete combustion of fossil fuels, primarily from vehicles but also from other combustion processes. This gas is odorless and tasteless, and it inhibits the ability of oxygen to move throughout the human body and can contribute to heart disease. In concentrated quantities it can and does cause death.

Nitrogen oxides are collectively called NOx for short. Two types are nitric oxide (NO) and nitrogen dioxide (NO_2), which are formed during the combustion of fossil fuel at high temperatures, for example, in transportation (cars etc.) and in power plants. Nitrogen oxides are thought to contribute to heart and lung problems, including asthma, may cause cancer, and lower the body's resistance to respiratory infection. They are also a component in the formation of low-level ozone and acid rain.

Sulfur oxides (SOx) are gases that typically come from burning fuels containing sulfur, mostly coal but some from oil. Sources include power plants, petroleum refineries, paper mills and smelters. Sulfur dioxide (SO_2) causes respiratory problems, such as aggravating asthma, bronchitis and emphysema. It is also a component of acid ran.

Secondary pollutants are caused by reactions involving primary pollutants and are also called photochemical oxidants. This is because the sun supplies the energy for the reactions to take place, and this also causes photochemical smog. The reaction of VOCs and NOx with sunlight forms low-level ozone (O_3), peroxyacetyl nitrate (PANS), aldehydes and ketones. These too are troublesome for the eyes and lungs. Ozone can permanently scar the lungs, impairing lung function, and it is thought to be the cause of hospitalization for thousands of asthmatics each year.

Sometimes called a secondary pollutant, acid deposition was a problem in the 1850s and was well established as a problem by the 1940s. It is a worldwide problem, and it can be wet, or it can settle out of the atmosphere as dry particles. It is more concentrated in crowded industrial areas such as the eastern half of the U.S. Any precipitation—rain, snow, fog or mist—can be more acidic than normal due to the burning of fossil fuels. The most common term for this is acid precipitation or acid rain, which refers to just that, higher than normal acidity in rain. Precipitation that is 10 times more acidic than normal has become common, and there are cases of it being 1,000 times more acidic than normal.

Sulfuric acid and nitric acid are the main components of acid precipitation, dropping out of the sky in dilute form caused by complex chemical reactions in the atmosphere. Sulfuric acid comes mostly from burning coal (in power plants), whereas nitric acid comes substantially from tail pipe (vehicle) emissions, although these are not the only sources. As with all air pollution we also must look to aircraft, ocean going vessels, home heating and hot water, garden equipment, recreational vehicles—and anyplace else where fossil fuel derivatives are burned. Acid precipitation can be carried far away from its source, so beautiful, pristine forests and lakes are severely damaged.

The link to power plants was well known by the 1980s, but in the U.S. politicians and corporate executives from the fossil fuel and electric power industries blocked all progress to reduce acid deposition. By the 1990s legislation was passed to reduce acid-causing emissions and amendments have since been added, however political delays hampered progress for many years. In Europe, legislation has also been passed and overall acid deposition has been reduced. However, all power plants do not have the equipment to remove sulfur and while many are upgrading, many do not work efficiently enough, at least in part due to improper maintenance. In the U.S. emission allowances have also been instituted. Of course, many areas of the world do not clean up their emissions and burning coal is not the only source of acid deposition. The problem still persists and likely will for many years.

Much of the time the damage from acid precipitation and air pollution is manifested in plant life becoming weakened and therefore more susceptible to drought, disease and pests. As acid precipitation alters soil chemistry forests all over the world are damaged, and the effects of air pollution (especially low-level ozone) make matters worse. Damage to leaves and bark and reduced photosynthesis and growth are common, and root systems sometimes do not fully develop. Acidification of soils damages trees and other plant life by diminishing nutrient uptake, and by damaging vital soil fungi that have symbiotic relationships with many plants. Ground-water supplies also are polluted, as aluminum is leached out of the soil by acids. Birds living in areas of high acidity have been found to be more likely to lay eggs with fragile shells that break before hatching. This is because less calcium is passed up the food chain to the bird since calcium is being leached out of the soil by acid. Acid precipitation adds to habitat destruction and so to species extinction.

In some areas, as much as fifty percent of Germany's southwestern Black Forest has been damaged or destroyed by acid precipitation, and similar problems are evident across Europe. Since the 1970s in the mountains of the northeastern U.S. about half of the red spruce trees have died. Sugar maples in the eastern sections of the U.S. and Canada are also dying, and in some forests over 50 percent of the trees are damaged. From Canada to Maine, South Carolina to Georgia, especially at high elevations, trees are in decline. Specific studies in Vermont and North Carolina show areas of severe tree decline. On North Carolina's Mount Mitchell almost all the Fraser fir and red spruce above 6,000 feet are in decline. In the western U.S. similar problems plague forests. There have been some improvements in some areas but as long as we keep burning fossil fuels, the problem will persist.

Plants are very sensitive to air pollution as well as acid precipitation, and photochemical pollution, especially ozone, does much of the damage. Crop damage from low-level ozone pollution is estimated at between two billion and seven billion dollars a year, and in California alone it may be over one billion.

Entire fish populations in thousands of lakes all over the world, including Canada and the U.S., have been negatively affected, and countless numbers of fish have been killed. Even Norway and Sweden have been catastrophically affected. In Sweden many lakes are so acidified that many fish species are gone. In New York's Adirondack Mountains above 3,000 feet acidification has also eliminated many fish species. Some lakes have been found with no fish. What about the others species dependant on the fish? Also, when waters become more acidic, mercury can accumulate more easily in fish. Some states near the Great Lakes have recommended not eating the lake fish for this reason.

Acid deposition also does extreme damage to human structures, monuments and equipment. In the U.S. repairs run into billions of dollars every year, with similar problems in Europe and the rest of the world. Bridges, roads, steel, concrete, paint, rubber, limestone and marble are all damaged by acid rain and air pollution. Stained glass windows in the gothic cathedral in Cologne are deteriorating. The Lincoln Memorial, Washington Monument, medieval cathedrals in Europe, the

Coliseum in Rome, frescoes and statues of Florence, the Parthenon in Athens and the Taj Mahal all show signs of deterioration.

Reduced visibility from pollution is everywhere, including once untouched areas such as Grand Canyon National Park. Visibility there was once typically 180 miles. Now sometimes it is only 12 miles across areas of the canyon. Most of the U.S. now has pollution spread across it. The air was clear at one time, but in areas of the U.S. visibility has been reduced by as much as 80 percent. Low visibility of 9 or 10 miles has become typical, whereas if we stopped polluting completely, visibility almost everywhere would jump to about 90 miles. The rest of the world faces similar circumstances.

The eastern half of the U.S. is not only home to the most acidic rain but also the most mercury pollution. Mercury is a neurotoxin. It devastates the central nervous system and the brain, and can cause kidney damage. Developing fetuses and young children are most at risk. In high enough quantities blindness, mental retardation and other neurological disabilities have been seen. Warnings about eating locally caught fresh-water fish have been issued by 45 states. Ocean fish, including marlin, shark, tuna, swordfish and lobster, also have been shown to contain high levels of mercury, and young children and pregnant women have been warned about consumption levels. In 2005 the U.S. National Institutes of Health estimated that 300,000 or more children born every year in the U.S. may have diminished intelligence or developmental problems because of mercury exposure in the womb. For people and wildlife, the most common mercury exposure is through eating fish.

Power plants that burn coal, as well as medical and municipal waste incinerators, generate most of the anthropogenic mercury in the atmosphere. The smelting of metals such as zinc, lead and copper, and the manufacturing of some chemicals also release mercury. This mercury finds its way into terrestrial and marine ecosystems. U.S. coal-burning power plants produce over 50 percent of the yearly mercury emissions in the U.S. Other industries that emit mercury have been regulated, but over 1,000 coal-burning power plants are not regulated adequately. In 2005 the Bush administration blocked an attempt to mandate control technology to dramatically reduce mercury emissions at these plants. Instead, a market driven mechanism allowing the industry to buy and sell pollution rights was instituted, the goal being to reduce mercury emissions by 70 percent by 2018. But the Congressional Research Service estimates that it is unlikely this reduction will take place before

2030. This issue is being reevaluated. In any case, mercury pollution is generated all over the world, as is most pollution, and it deposits all over the world.

Indoor air pollution can be much worse than outdoor air pollution. The air from outside is also inside our homes and other buildings. It mixes with what we add to it, making indoor air sometimes very unhealthy. Mold; household cleaners and other chemicals; perfumes; synthetic chemicals emitted from carpets, plastics, wall coverings and building materials; cigarette smoke; methane from stoves and combustion of methane and oil from stoves and heating systems; evaporation from paint and solvents, and radon can all be in the home. The pollution from lawn equipment is very problematic because it can be so constant and concentrated around the home and so enters into the home.

Sick building syndrome has become common, especially in buildings with windows that do not open and in which the air is run through a recirculation system. The EPA estimates that sick building syndrome costs about 60 billion a year in reduced productivity, absenteeism and medical expenses. Clearly, a home or building should be routinely aired out.

The World Health Organization estimates that every year illnesses related to air pollution prematurely kill five to six million people. As already mentioned, in the U.S. as many as 60,000 people a year die prematurely from air pollution. Asthma, bronchitis, emphysema and lung and heart disease, including lung cancer and heart attacks, can be aggravated or even caused by air pollution. In the U.S. residents in the most polluted cities are more likely to die from some of these ailments than those living in the cities with cleaner air. Every year in the U.S. thousands of children have asthma attacks that are induced by air pollution. It has been estimated that 250,000 people in the U.S. suffer from bronchitis and emphysema related to pollution. In the U.S. 10,000 heart disease fatalities may be related to particulate (PM) air pollution. Put simply, the more polluted an area is, the more people get sick or die.

In the U.S. about one half of cancers related to outdoor air pollution are linked to pollution from cars and other mobile sources. In seven U.S. cities—New York, Los Angeles, Houston, Chicago, Detroit, Milwaukee and Philadelphia—a four-year study showed that carbon monoxide (CO) (mostly from vehicles and industry) in the air was linked to hospitalization for congestive heart failure. CO causes a reduction in oxygen transport in the blood, headaches, fatigue, and at high levels, death.

Low-level ozone from burning gasoline causes asthma attacks and immune-system suppression and has been linked to heart defects in newborns, as well as to deaths from respiratory and cardiovascular problems.

Children are much more susceptible to air pollution than are adults. Their lungs are still developing, and proportionally they breathe in more air, about two times as much per pound of body weight. This makes a child more likely to have pollution-related health problems in later life. Studies have shown lung damage in 80 percent of 100 Los Angeles children who died of unrelated reasons. In a southern California study, five thousand children who were chronically exposed to air pollution were found to be more likely to develop asthma if they participated in sports than those who did not participate. This study also showed that in the more polluted areas—those with higher levels of PM, NOx, and acid vapor—some children have impaired lung growth.

The world's atmosphere is all one, and many things mix within it. Corrosive gases, smoke, dust, haze, odors, and toxic compounds are present everywhere to some degree, even in the remote wilderness, and clearly major cities can be worse than rural areas. Even over the Arctic, a brown pollution haze forms each spring. In 1998 a satellite tracked a massive cloud of particulate matter across the Pacific Ocean, and when it reached the U.S. it was found to contained zinc, copper, lead and arsenic from Chinese smelters. Noise pollution too, is carried by the atmosphere and has become a serious problem for health.

The EPA estimates that in the U.S. alone, about 147 million metric tons of air pollution were released to the atmosphere in 2005. Since then the amounts are similar. This does not include CO_2, which only recently was finally classified as an air pollutant in the U.S. Worldwide, air pollutant emissions amount to about two billion metric tons per year, again excluding CO_2. Worldwide, yearly CO_2 emissions from human activities are over 26 billion metric tons.

14
WATER

Some water molecules do escape the atmosphere (volcanoes replace some of this), but practically speaking the world will not run out of water. Gravity holds the planet's water in place as it constantly changes form, moving in all directions between its liquid, gas and solid states. Always on the move, it condenses, evaporates, freezes and melts in a perpetual round trip of change. However, what we sometimes do to water is pollute it, causing us to no longer be able to use it, and we also cause it to move from places where we have access to it, to places where we do not.

Already mentioned, the hydrologic cycle is the water cycle, or simply put, how water moves around the planet. Water is life-giving. Almost nothing biological, no plant or animal, can survive without it. Our food and drink, our bodies, human systems of manufacturing and entire economies all depend on it. Its constant movement absorbs and releases heat and so is vital to livable climate and weather. The circulation of water from a water body to the atmosphere, to the land and back to the water body again is the basis of the hydrologic cycle.

Water evaporates from all surface water—the most influential body of which is the ocean—into the atmosphere, forms clouds, and becomes precipitation. Rain, snow, hail, sleet or even fog bring the fresh water we need back down again to the land and oceans, and the process keeps on going, like a circular conveyor belt. Any form of evaporation and condensation is part of the process. Soils too, contain water that evaporates into the atmosphere, and so does plant life. As already mentioned, plants transpire as part of photosynthesis, and new water is formed (the original water (H_2O) is broken up to supply the atmosphere with oxygen) to be released as water vapor from pores (stomata) on plant parts, a leaf for example, the same place the oxygen is given off. Plant life, especially forests, contributes gigantic quantities of water to the atmosphere. The evaporation and transpiration of water is called evapotranspiration.

When precipitation—rain for example—comes down and hits the land, some runs off into bays and rivers and from there back to the ocean, while some water runs off directly back to the ocean. Some soaks into the ground and recharges underground aquifers, where water is trapped, sometimes for hundreds or thousands of years. We tap into this underground water source. Water stored in aquifers will sometimes surface again in lakes, ponds or rivers to make its way back to the oceans, either directly or through evaporation. It may even find an underground path back to the ocean.

Ice and snow melts, or sublimates (goes directly to the vapor state) and rejoins the hydrologic cycle in similar ways, even though in places such as the Arctic and Antarctic, for example, some ice and snow may not melt for very long time periods. Still, water in any form is part of the cycle. The amount of water in volume that joins the atmosphere every year is estimated at 100,000 cubic miles. Only about 25 percent of this falls on land as precipitation; the rest falls on the oceans.

If I stand in my backyard and spray water into the air from a garden-hose am I wasting water? (I certainly am wasting energy.) The answer is yes, and no. Some of the water will evaporate from the water stream and off the lawn, going back to the atmosphere to rejoin the hydrologic cycle. If this water rains back down locally to recharge my aquifer or above-ground water supply, it is not wasted; neither is the water from the hose that hits the ground and seeps back down. But, how long will it take for all this to recharge my aquifer? Possibly longer than I can wait, and the aquifer may not recharge at all if enough land is covered by human structures.

The water that has evaporated is wasted as far as I am concerned if it does not rain back down locally, and it may not. It may rain down over an ocean or some other state or country. If enough water is used up, a town—or a large area in the state of Nevada for example—can have catastrophic water shortages, especially if there is an ongoing drought.

However, on a worldwide scale, I did not waste any water because, as I said, the total quantity of water is fixed; I only redistributed the water—although doing so might cause my death by dehydration, as happens in places like Africa. Someone else might get to use the water, or it might go back to the oceans with no one having a chance to use it during its trip.

Water distribution on earth is not in favor of human beings, considering that about 97 percent of it is salty seawater: we can't drink that, and neither can terrestrial plant life. About 2.5 percent is fresh water that we can use, but most of this is frozen in ice caps and glaciers. A small percentage of fresh water at any given time is in the atmosphere and in the soil, where it is sometimes not accessible to humans. The remainder is only about one-half of one percent of the total water on the planet available for human consumption, and this is found mostly as groundwater (aquifers, of which there are different types), with a small portion of this half-percent found in surface water bodies such as lakes and rivers.

As would be expected, as population and economies grow, the demand for fresh water also is rapidly growing, and to make matters worse, in many areas per capita water use is increasing. By 2025 over 30 percent of humans may not have adequate water for irrigation and drinking. Even today, the problem is severe. The World Health Organization estimates that improper sanitation, poor water quality and low water supplies are responsible for the majority of human illnesses. Industrial wastes and sewage contaminate much of the water people drink in many areas of the world. Worldwide, over 2.5 billion people do not have the proper means for disposal of domestic wastewater and fecal waste, while about 1.5 billion people do not have safe drinking water. Every year, all over the world, water-related sickness affects over 250 million people and about five million die. Improvements are being made, but poverty and population pressures impede progress.

In developing nations water use might be as low as a gallon a day per person, while in developed nations such as the U.S. it is measured in hundreds of gallons per day, if we include industrial, agricultural and personal consumption. (One toilet flush can use more water than a poor person in a developing nation uses all day for everything.) The least amount of water consumed is municipal and domestic which represents about 10 percent of the worldwide total. Industry consumes about 20 percent, and irrigation for agriculture takes about 70 percent.

For humans, the source of fresh water is surface water and aquifers. Surface water means lakes, rivers, ponds and human-made reservoirs such as the water behind a dam. These are technically renewable sources, thanks to the hydrologic cycle, but as we change the hydrologic cycle through global warming this too is changing. Above ground supplies can easily dry up, and this is happening in parts of the world. (This change also goes the other way with flooding.) Overuse adds to the problem, and sometimes as much as 70 percent of a surface water body can be depleted. This is not only dangerous for humans but has environmental consequences as well. Plants, birds, fish and other animals are affected from the lack of water, and pollution and salt tend to concentrate as water diminishes.

Aquifers in the U.S. supply about 50 percent of our drinking water and about 40 percent of water for irrigation. Aquifers vary depending on soil types (sand, gravel, rock, clay, etc.), the porosity and permeability of the ground, and the depth of the water, which can be a few feet from the surface or several thousand feet down; it's typically measured in hundreds of feet. It's not a river down there but rather tightly packed material with water squeezed between grains under great pressure. (Occasionally there are voids, such as caves that are filled with water.)

Aquifers would be renewable if left alone to naturally recharge from precipitation and sometimes surface water. However, much of the time humans draw underground water so fast that it can never recharge in the human time frame. All over the U.S. (in Arizona, and California, and from North Dakota south to Texas, for example) and all over the world, groundwater levels have dropped, especially since the 1950s as population exploded. Las Vegas is an example. The aquifer under this desert city is almost depleted. Now that the desert floor is covered with hotels, concrete, roads, houses, golf courses, etc., the little rain that does come down washes off or evaporates more than it would have. The aquifer may never naturally recharge in time to be useful for humans, especially considering that many aquifers take hundreds or thousands of years to charge—through water seeping down into the ground—in the first place. Modern structures essentially kill many aquifers in terms of human use by blocking recharge. Overdrawing of aquifers also can and does lead to ground subsidence, where large areas of the surface actually sink.

The Ogallala aquifer is the largest groundwater source in the world. It underlies about eight states from South Dakota down to Texas. In parts of this aquifer because of agriculture, raising cattle and other uses, water is being removed 40 times faster than precipitation can put it back and water tables have dropped by as much as 100 feet. (A water table is the top of an unconfined aquifer.) During dry periods this has caused serious problems for farmers. The water in the Ogallala is not distributed evenly, but water experts believe that eventually there will be water shortages in all areas of this aquifer. Similar problems exist all over the world. In addition to shortages, along coastal areas saltwater intrusion can ruin a well as it is over-pumped and draws in nearby saltwater.

In the U.S. not only are underground water supplies being strained, but surface water problems are cropping up too. Overall, the U.S. has had a good water supply, especially when compared to other countries, but our population is increasing, we habitually waste water, and droughts are not uncommon. The Southwest and the West—semiarid or arid regions—are experiencing serious water problems, and in the West a six-year drought continues. Ninety percent of U.S. industrial water use goes to coal, petroleum, metals, chemical, paper and pulp, and food processing, and along with development, industrial, commercial and domestic water use has been on a non-stop increase. Las Vegas is a perfect example, as it has been the fastest growing city in the U.S. The western drought is believed to be a natural cycle that may last 20 or more years, and global warming is intensifying the problem. Further, some scientists believe that the drought may no longer be simply a drought but rather a permanent climate change—a new climate of much less water—that we must accept. Less water and more evaporation! Water levels are down all over the U.S. and the world.

If the climate doesn't give the western U.S. a reprieve, then serious trouble is very near, and eventually there will not be enough water to support the affected areas. Lake Mead, Nevada, behind Hoover Dam, is so low that marinas have been left surrounded by dry land—and power boats should not be on a drinking water reservoir to begin with! Lake Mead is about half empty, and some models show that not only will it never be full again, it is going lower. Hopefully the models are wrong because it supplies almost all of Las Vegas's water—water for 1.8 million people.

Lake Powell in Utah, is at about 50 percent capacity, the lowest level since 1973, and docks there have been moved to chase the water. There are plans to find and divert water—for example, running a pipeline from northern Nevada's rural areas—but this will stress these other areas' water supply. Ranchers and farmers in these rural areas—who supply cities with food grown with that water—are ready to fight the cities in court over their water. Gee, maybe the cities should stop expanding.

The largest body of fresh water in the world is Lake Superior, and it supplies millions of people with water. During September 2007, the water in the lake was at its lowest level since 1860, when record keeping began. The water level was lower than the record set during the Dust Bowl drought days of the early 1900s.

The Colorado River supplies Los Angeles, Las Vegas, Salt Lake City, Denver, Phoenix, San Diego and Albuquerque with water: over 30 million people in the U.S. rely on it, and so Mexico is now losing their water supply from the Colorado. Global warming may eliminate 25 percent or more of snowpack over the next 50 years, some say much sooner, so the whole cycle of precipitation and snowmelt that feeds the Colorado is changing while water demand is dramatically increasing. Snowpack in some areas is at its lowest level in 20 years. Agriculture is threatened, along with electricity production from hydropower at 11 dams out of the 49 along the Colorado. Pollution and the river becoming salty (in part through evaporation) are also serious problems.

Mexico faces catastrophic water shortages as several key aquifers are rapidly diminishing. There have been riots in Pakistan because of a drought from the 1990s into the early 2000s that caused food problems and additional poverty. Over 30 percent of the wells in Beijing, China, have gone dry. Six hundred million people face water problems across the North China Plain as the water table falls 6 to 10 feet per year. Over a 15-year period, China's Yellow River has run dry far inland before reaching the Yellow Sea. Most of China has water problems (and extreme water pollution). In areas of India water rationing has been instituted and local water in over 1,000 villages does not exist anymore.

Worldwide there are over 200 watersheds (land areas that distribute water to a river system or stream) that are shared by two or more nations. Luxembourg, the Netherlands, France, Germany and Switzerland share the Rhine River watershed. When Switzerland in 1986 accidentally dumped 30 tons of chemicals into the Rhine, killed over 500,000 fish and polluted water supplies, a lot of people were not happy.

Part of the former Soviet Union, the Aral Sea—a fresh water lake, once the world's fourth largest—has been reduced by 80 percent of its volume, because of water diversion projects to irrigate crops in surrounding areas. Fish species have collapsed, and 35 million people around the lake have been affected with health problems, including high death rates from respiratory illness possibly caused by wind blown dust storms from the dried up shoreline. (Hundreds of tons of anthrax-related biological warfare products were buried on an island in the Aral. It turned out not to be a good idea. Problems are arising.)

So many watersheds and underground aquifers are shared country to country that problems are bound to arise, including possibly armed conflicts. Vietnam, Laos and Thailand share the Mekong River basin. The Danube River supplies both Hungary and Slovakia. India and Pakistan share the Indus River basin. Syria, Iraq and Turkey rely on water from the Tigris-Euphrates river basin. Jordon, Israel, the Gaza Strip and the West Bank all rely on the Jordon River. The Nile River basin in northeastern Africa supplies 10 nations, including Ethiopia, Egypt and the Sudan. Some of these areas are already in conflict, and threats have been made over water use, as one area takes water from another area and supplies dwindle. Even in the U.S. states are starting to argue over water, for example, Florida and Georgia.

Conservation of water in agricultural, industrial and domestic uses would go a long way toward helping the problem of water shortages. As with gasoline, the U.S. could dramatically cut water consumption just by changing habits such as not washing a car as much, not playing with the garden hose as much, not letting water run so much and repairing leaks. In the U.S. golf courses, lawns and parks totaling about 75 million acres take more water—not to mention pesticides and fertilizer, which then pollute aquifers—than any other land. The average American uses over 75 gallons of water a day—about 30,000 gallons a year—for home use. If other indirect uses are figured in—such as irrigation for food—then the number runs over 1,200 gallons per day, or over 438,000 gallons per year. Consider this: a new car requires about 10,000 gallons of water to manufacture; a glass of milk over 50 gallons to produce, and a leaky faucet or toilet bowl can easily waste over 1,000 gallons a year. Let your imagination run, and you will find a lot of ways to save water.

Much has been done with all kinds of water-saving devices, from low-use toilets and flow restrictors on faucets and showers, to the treating and recycling of waste water for reuse, to new methods of irrigation, but more needs to be done overall. As much as 20 percent of piped water in U.S. cities is lost to leaks before it gets where it is going. This needs to be addressed much more than it has been.

Desalinization is a process in which saltwater has the salt removed from it. There are many small desalinization plants around the U.S. that are designed to remove salt from water that is not as salty as seawater, such as brackish groundwater. Large-scale seawater desalinization plants are another story. There are only a few. The process works, and it has the potential to create a very large freshwater supply. However, the process is very expensive from a dollar standpoint as well as an energy one. Very large amounts of energy would be needed to desalinate the gigantic quantities of water we use. If seawater desalination is to be used for irrigation in agriculture, then it becomes even more expensive because of the large quantities of water needed. And there are other problems that sometimes occur with seawater desalinization, depending on how it is done; levels of important minerals in water that crops need can be inadvertently reduced to the detriment of crops, creating the need to replace the minerals.

Two processes are generally used for desalination: reverse osmosis and distillation. Reverse osmosis forces the saltwater through a membrane that removes most of the salt—it is the less expensive of the two processes. Distillation leaves the salt behind by heating and evaporating the water. Saudi Arabia—where it is cost effective—obtains over half its water from about 28 seawater desalination plants. Other areas of the world, such as the Middle East and North Africa use desalination extensively. However, only about 1 percent of the world's drinking water comes from seawater desalination.

The average middle-class American neighborhood might not deal well with the dollar cost of desalinated seawater, and additional environmental damage would be sustained from building and maintaining the plants, as well as from the energy used to desalinate (air pollution, etc.) and the massive quantities of salt that would have to be disposed of. Desalination becomes less expensive when waste heat from fossil-fuel power-plants is used for the process—as is the case in the Middle East—but this assumes that we will keep burning fossil fuels for our energy needs. Further, desalination plants are better located near coastlines where many power plants are not located. Near Yuma, Arizona, in 1992 a large desalination plant was built along the Colorado River, but it was so expensive to operate that it was shut down in less than one year. In 2003 in Tampa, Florida, a seawater desalination plant started up, and the water produced costs about 2.5 times normal water supplies. The plant ran inconsistently, but as of 2007-2008 it should be up and running on a permanent basis. However this plant, the countries largest, produces only about 25 million gallons of fresh water a day, which supplies only about 10 percent of the area's water needs. The U.S. uses over 125 trillion gallons of water a year! The challenge for seawater desalinization is enormous, and it probably will not be able to solve the problem completely, especially considering our population and demand for resources keep growing—which I repeat, must stop. With or without seawater desalination, water is going to get more expensive. Conservation and water recycling must take a front seat.

Technology is advancing, and if solar power is used, desalination may someday become more common and less expensive. Regardless, the developing nations that can least afford it have some of the worst water problems. It takes time and lots of money to construct the infrastructure—plants, pipes, etc. The U.S. has not seriously begun the process but will as water problems worsen. California is proposing to build desalination plants, and other states are looking at the issue. However, will we prepare in time—"in time" is the key—before water problems induced naturally and from global warming overwhelm us, and will inexpensive desalination come in time? It can easily take five years to approve and build a large plant. At minimum, hundreds of huge plants will be needed, maybe thousands, depending on how serious water problems become. If plants are built on the scale of the Tampa plant, then about 15,000 plants would be needed to take care of current U.S. water needs. And, if seawater desalination is ever fully established, will we then euphorically run off in triumph over nature and blindly grow our population and economies some more?

Water shortage is not the only problem. Water pollution is a worldwide dilemma, especially in developing countries, and the types and levels of pollution vary. In the U.S. drinking water is generally of high quality, and the system of distribution is excellent. However, there are still problems, and contaminated wells do get shut down. In 2002 the U.S. EPA stated that 51 percent of our estuaries (bay like areas, with a mix of fresh and salt water), 39 percent of our rivers, and 45 percent of our lakes were so polluted that they could not be used for fishing, swimming or drinking. Oceans and water bodies all over the world are polluted. Some dolphins and whales are so full of toxic compounds such as PCBs, that the animals can be considered toxic waste. Many fish and mammals on top of the food chain have biologically accumulated dangerous levels of pollution. Eating fish can be unhealthy.

Typically, water pollution is classified as either point source or nonpoint source. Point-source pollution, such as from sewage treatment plants or specific industries, is more easy to identify and control. Nonpoint is more difficult to find and stop because it comes from larger areas. Examples are agricultural runoff, construction site erosion, mining waste and residential lawns. Nonpoint pollution is a major problem in the U.S. that has not been adequately addressed.

Agriculture is the cause of most surface-water pollution in the U.S., and it contributes to over 70 percent of river pollution. Fertilizer runoff, pesticides and animal waste are the problems. Many of these things also pollute groundwater. One study of specific water supplies showed that about 50 percent of groundwater samples and 95 percent of stream and river samples contained at least one pesticide, while 10 percent of stream samples showed 10 or more pesticides.

Over 150 million Americans—at least half the country's population—live near seacoasts or the Great Lakes. Forests and wetlands have been destroyed as homes, strip malls, resorts, industrial and commercial enterprises replace them. Concrete and asphalt are the new ground cover. Runoff of animal and human waste (fecal matter) and improperly designed septic systems allow fecal bacteria and microbes that cause disease to spread into groundwater and surface water, endangering the health of marine life and people. Improperly designed sanitation systems, in which storm drains mix with sewer lines, cause overflows when heavy rains hit, dumping human waste into water bodies. In 2004 over 1,200 beaches—freshwater and ocean—experienced almost 20,000 days of closures and pollution advisories. High counts of fecal bacteria in the waters were the reason for 85 percent of the problem. Eating shellfish can cause disease and some shellfish beds are closed during and after heavy rainfalls as pollution increases from runoff.

The EPA has stated that these types of pollution are the main cause for U.S. water quality problems. There are ways to clean up the mess, and things are being done, but as growth and population continue to increase, the cleanup becomes more expensive and difficult.

The U.S. Great Lakes contain vast amounts of fresh water—Lake Superior has already been mentioned. About 20 percent of the world's fresh surface water is held by the five Great Lakes. Drinking water for about 38 million people comes from the Great Lakes, and about the same number of people live in the lakes' watershed. But for several hundred years the lakes have also been the receptacle for fertilizers, sewage, industrial waste and garbage, as well as many other pollutants, today including lawn chemicals. In the 1960's the pollution—especially from toxic chemicals, some carcinogenic—was so severe that serious deformities of animals in and around the lakes were common. Massive fish die-offs also became common, and it wasn't considered safe to eat fish from the lakes or to swim in any of them—although some were more polluted than others. Today the lakes are improved but still dangerously polluted. An array of pollutants is found in the lakes, including mercury. Fish, birds and other animals still show internal levels of PCBs and other chemicals, and reproductive failures and developmental and behavioral problems have been seen. Advisories are issued about eating lake fish, and there are potential health problems from doing so. Breast milk of women living near the lakes still shows levels of DDT. Continued development still damages shorelines and adds to pollution, but it is said that the drinking water from the lakes is safe.

In 1993 in Milwaukee drinking water was contaminated with Cryptosporidium protozoa from fecal waste, and 400,000 people became ill, while more than 100 died. Cattle feedlot runoff in 1999 in Washington sickened over 100 people, and in 2000 in Ontario, 2,300 people became ill and some died from Escherichia coli (E. coli) from the same cause. E. coli is very common in sewage—human and animal feces—and is deadly.

Polluted water can transport bacteria, parasites and disease. Some examples are cryptosporidium, ancylostomiasis, schistosomiasis, infectious hepatitis, poliomyelitis, typhoid, enteritis, dysentery, cholera and E. coli. The resulting infections range from flu-like symptoms to diarrhea and vomiting, internal hemorrhaging, organ damage and death.

Livestock waste from the factory farming (CAFO) of cattle (meat and dairy), pigs and poultry is another major problem as these industries have become gigantic. The disposal of vast quantities of manure by spraying or spreading leads to runoff into water bodies, threatening people and the environment. In the U.S. livestock feces and urine are produced in quantities that are over 20 times the amount of human feces and urine produced, and none of this is treated and processed as most human waste is.

Eutrophication of water bodies takes place when too many nutrients enter the water. Animal and human waste and fertilizer runoff from residential and agricultural land, as well as organic debris, cause excessive growth of algae. Nitrates and phosphates from these sources overfertilize a system. Water becomes cloudy, blocking out sunlight, and algae and aquatic plants begin to die off. Bacteria do their job of breaking down dead organic matter, but this depletes the water of oxygen. The lack of oxygen can kill most marine life. Marine life that can move avoid the area, but bottom dwellers such as clams die from lack of oxygen. This is a problem all over the world, and not a small one; it has been reported in over 400 coastal areas around the world and in many inland water bodies. Additionally, global warming may be adding to the problem by heating oceans and so reducing oxygen levels, and by causing wind pattern changes in coastal areas. Periodically the Gulf of Mexico has a dead zone caused mostly by runoff from agricultural and livestock operations in states along the Mississippi River. This dead zone can last from March to September every year, and it runs all the way from the bottom to near the surface. Its size has been equated to roughly that of Massachusetts, and it is one of the world's largest dead zones and affects vast quantities of marine life.

Another concern is sediment pollution, which is caused by the erosion of soil. Soils exposed by cutting down trees, agriculture, overgrazing domestic animals, construction and strip mining all contribute significantly to the problem. Sediment (soil particles) in water bodies reduces light penetration, contributing to eutrophication. Sediments also settle to the bottom and cover organisms such as shellfish beds and coral reefs, seriously damaging or killing them, which in turn can kill off fish populations. Toxic chemicals and disease-causing microbes are also carried by sediments, affecting people, marine plants and animals. Some waterways are seriously polluted by toxic sediments.

Chemical pollution from human activities is almost everywhere. Industries still dump waste directly into waterways—although this is regulated in developed nations—while agriculture, livestock operations, and residential and commercial lawns as well as landfills cause runoff and seepage of chemicals and other pollutants. Both surface water and groundwater are affected. Even traces of antibiotics, pain killers, hormones from birth control pills, insect repellants and fragrances have been detected in water samples. And some pollutants leak into our drinking water through broken pipes, undetected, after treated water leaves the water plant.

There are a wide range of chemicals creating problems, from solvents to pesticides, household products, industrial chemicals, components of gasoline, fumigants and plastic compounds, to name a few from the thousands of chemicals used every year. Some chemicals found in polluted water are benzene, chloroform, ethylene dibromide, polychlorinated biphenyls (PCBs), trichloroethylene, vinyl chloride, dioxins, carbon tetrachloride and aldicarb. The heavy metals lead and mercury (both dangerous neurotoxins) have also been found in drinking water. The effects of these things if ingested are wide ranging and of course depend on dosage. In very small quantities, parts per billion for example, there may not be an effect on biological systems; however, as previously mentioned, over time some pollutants bioaccumulate in the system. In addition, quantities are not always small, and some pollutants such as lead, mercury, PCBs and dioxin are dangerous even in low quantities. Effects of different chemicals and heavy metals can range from nervous-system damage to hypertension, miscarriages, learning disabilities, brain disorders, cancer, liver and kidney damage, blood disorders, reproductive and immune system damage and even endocrine system (hormone) problems. Many chemicals that are used by society are not needed and are used in quantities that are far too great.

Developing nations often dump sewage directly into harbors and rivers. In Africa it is common for disease to spread through drinking water. Programs like the one in Kwale, Kenya, where deadly diarrhea and cholera outbreaks have taken place, have attempted to solve drinking water problems. Gaining access to ground water instead of using surface water has helped, but droughts have persisted, and cholera from pollution continues.

Even in Italy, the huge Po River has been polluted by industry, sewage, agriculture, erosion and sediments. Toxic chemicals and dangerous microbes have been a serious problem there. Fishing, as well as tourism, has been affected, and beaches have been closed due to pollution, yet drinking water is still taken from this river.

There are pollution problems all over South America. Representative of this is Lake Maracaibo, Venezuela. Run-off from industries and farms, sewage and oil and gas production—oil routinely leaks into the lake—have devastated the lake. Sewage treatment plants have been built but much more needs to be done.

The famous Ganges River in India is extremely polluted. Industrial waste, sewage, chemicals, and even (because of religious beliefs) human remains have been put into the river, and all have had a devastating effect. Over 350 million people live in the river's basin, and many wash their clothes and themselves in the river. Millions of dollars have been spent on cleanup projects, but problems persist.

One of the most unbelievable cases of drinking-water pollution comes from Bangladesh, where a project to help the people obtain clean drinking water resulted in tragedy. A project was put in place to install over two million hand-pumped wells throughout the area. The groundwater was assumed to be safe but turned out to have very high levels of naturally occurring arsenic in it, a problem that has occurred in other parts of the world, including the western U.S. Exact numbers are not known for sure, but as many as 70 million people in Bangladesh may be affected. Arsenic poisoning—which causes skin lesions and other debilitating effects—is happening to tens of thousands of people in Bangladesh. Arsenic ingestion can and does lead to cancer and death.

Groundwater pollution is a problem everywhere, even in the U.S. As already mentioned, about half the drinking water in the U.S. comes from underground, drawn out by wells. Agriculture, landfills, lawns and golf courses all contribute chemical pesticides, fertilizers and organic matter to the pollution problem. Underground storage tanks holding gasoline, oil for home heating and different hydrocarbons for industry are also a major problem. Many of them leak, and over 250,000 holding gasoline at service stations are potential problems.

In rural areas 80 percent of the residents use wells. In some rural settings fertilizer from farm operations can be a hazard, as nitrates from the fertilizer seep into the ground. (Nitrates are particularly dangerous to newborns.) Nitrate pollution is also a problem in suburban areas because of lawns. Most at risk are wells drilled down to 100 feet or less. Privately owned wells are the responsibility of the homeowner and should be tested routinely, as they are susceptible to all forms of groundwater pollution, natural and unnatural.

When rainwater washes into and over an area such as a road or highway, many pollutants are picked up, such as copper, hydrocarbons, oil, lead, organic waste, zinc, acids, asbestos, salts, fecal matter and construction residue. Combination sewer systems —those mixing human waste and runoff from urban storm sewers—send the mix to sewage treatment plants. When the systems are overloaded during heavy rains or snows, the untreated overflow sometimes ends up in water bodies. The problem is being worked on, and some areas do not use combined systems, but even in the U.S. many cites have combined systems, and yearly over one trillion gallons of overflow end up dumped into waterways.

In developed countries such as the U.S., wastewater—everything that goes down a drain or toilet in a home or business (excluding septic systems)—is treated in plants. Primary treatment removes suspended particles, while secondary treatment biologically decomposes organic material using bacteria. Tertiary treatment uses additional filtering means and can make the water drinkable. About 10 percent of the U.S. uses only primary treatment and about 65 percent uses primary and secondary treatment. Tertiary treatment is expensive and is used by about 25 percent of the population.

Primary and secondary treatment leave behind very large quantities of sludge—solid material. If it is not too polluted with heavy metals and chemicals, it is sometimes used as fertilizer. It is also incinerated (causing air pollution) or dumped in landfills (contributing to habitat destruction and sometimes groundwater pollution). It used to be dumped in the ocean, but that practice was banned in the U.S. Thousands of treatment plants around the country run full time, and every day thousands of trucks continually move the sludge. The energy consumed and pollution generated by the process is enormous. Further, primary and secondary treatment do not remove pollutants such as heavy metals, chemicals and viruses. Ultimately the treated water is dumped into lakes, rivers or directly to the ocean.

On Long Island, New York, at one of several plants, about 70 million gallons of water a day is processed with primary and secondary treatment. It is then chlorinated and pumped into the ocean offshore from a popular beach. The water still has nitrogen and phosphorus nutrients in it, as well as some pollutants, and now also has chlorine—vast quantities of chlorine are used in the process. The ocean's abilities to dilute pollutants are certainly better than those of a lake or river. However, the oceans have been the world's dumping ground for far too long, and it is time to stop taking them for granted.

Further, this water originally came from an aquifer, or in some cases an above ground reservoir. The water should go directly back to where it came from in order to rejoin the hydrologic cycle in a way that keeps water quantities somewhat stable for the population. Remember the garden-hose analogy. However, when entire land areas are covered with roads, concrete, homes and commercial enterprises, many times it becomes too expensive, even impossible, to recharge an aquifer or a water reservoir that may be too far away. Therefore the processed water gets dumped elsewhere.

In rural areas of the country septic systems often are used for sewage disposal. Waste water is run underground into an enclosure where solids are naturally broken down by bacteria. Overflow of water runs to a leach field to be further broken down and purified by soils. The overflow water then evaporates or recharges the local aquifer. The systems work fine as long as they are not too densely located. Septic systems are the responsibility of the homeowner and must be maintained—and chemicals like bleach in waste water should be avoided.

About 60,000 facilities in the U.S. process water and send it to most of the population. The water is treated and checked for pollution. The final treatment usually entails adding chlorine to kill organisms that carry disease so that the water is safe to drink. Ozone or ultraviolet radiation (UV) treatment can also be used, and in much of Europe UV is used. Some type of treatment must be used or outbreaks of disease such as cholera or cryptosporidium would kill millions of people. However, there is a serious problem with chlorinating drinking water: it's not good to drink chlorine and the chlorine byproducts that form in the water. Chlorination has been linked to bladder and pancreatic cancer as well as to an increase in miscarriages.

Even with the Safe Drinking Water Act, the Clean Water Act and other regulations, pollution of water continues. More needs to be done to stop the pollution of surface and groundwater supplies. No matter how careful a facility is, sometimes pollutants get through. Water isn't always monitored properly, and as already mentioned, after it leaves the facility, it can then pick up pollutants. There is a very simple solution. Install a good-quality water filter on the home sink for drinking, cooking and ice making only. The better filters will remove chlorine and chlorine byproducts, as well as many other pollutants such as heavy metals, chemicals, and even cryptosporidium—which can become resistant to chlorine. Change the filter as per instructions, and for all practical purposes the problem is solved—unless, of course, your water supply someday dries up.

REFERENCES

Books

Ahrens, C. Donald. *Meteorology Today.* 7th ed. California: Thomson-Brooks/Cole, 2003.

Bowen, Mark. *Thin Ice.* New York: Henry Holt & Co, 2005.

Calvin, William H. *A Brain for All Seasons: Human Evolution & Abrupt Climate Change.* Chicago: The University of Chicago Press, 2002.

Carwardine, Mark, and Ken Watterson. *The Shark Watcher's Handbook.* Princeton & Oxford: Princeton University Press, 2002.

Chaisson, Eric, and Steve McMillan. *Astronomy.* 3rd ed. Upper Saddle River: Prentice Hall, 2001.

Cooke, Fred, et al. *The Encyclopedia of Animals: A Complete Visual Guide.* Berkley & Los Angeles: University of California Press, 2004.

Cunning, William P., Mary Ann Cunningham, and Barbara Woodworth Saigo. *Environmental Science, A Global Concern.* 9th ed. New York:

McGraw-Hill, 2007.

Cutnell, John D., and Kenneth W. Johnson. *Physics.* 5th ed. New York: John Wiley & Sons, 2001.

Dennis, Jack, et al. *The Nuclear Almanac.* Reading, Massachusetts: Addison-Wesley, 1984.

Diamond, Jared. *Collapse.* New York: Viking, 2005.

Eliot, Robert S., and Dennis L. Breo. *Is It Worth Dying For?* New York: Bantam Books, 1993.

Fetter, C.W. *Applied Hydrogeology.* 2nd ed. New York: Macmillan, 1988.

Firor, John, and Judith Jacobsen. *The Crowded Greenhouse: Population, Climate Change, and Creating a Sustainable World.* New Haven & London: Yale University Press, 2002.

Frohoff, Toni, Brenda Peterson, et al. *Between Species.* San Francisco: Sierra Club Books, 2003.

Garrett, Laurie. *The Coming Plague: Newly Emerging Diseases in a World Out of Balance.* New York: Farrar, Straus and Giroux, 1994.

Goodstein, David. *Out of Gas: The End of The Age of Oil.* New York: W.W. Norton & Company, 2004.

Hardy, John. *Climate Change.* W. Sussex, England: John Wiley & Sons, 2003.

Johnston, David Cay. *Perfectly Legal.* New York: Portfolio, 2003.

Kolbert, Elizabeth. *Field Notes from a Catastrophe.* New York: Bloomsbury, 2006.

Preston, Richard. *The Hot Zone.* New York: Random House, 1994.

Raven, Peter H., and George B. Johnson. *Biology.* 6th ed. New York: McGraw-Hill, 2002.

Raven, Peter H., and Linda R. Berg. *Environment.* 5th ed. New Jersey:

John Wiley & Sons, 2006.

Ristinen, Robert, and Jack J. Kraushaar. *Energy and the Environment.* 2nd ed. New Jersey: John Wiley & Sons, 2006.

Schaeffer, John. *Solar Living Source Book.* 12th ed. Hopland: Gaiam Real Goods, 2005.

Shnayerson, Michael, and Mark J. Plotkin. *The Killers Within: The Deadly Rise of Drug-Resistant Bacteria.* Boston, New York, London: Little, Brown and Company, 2002.

Stern, Kingsley R. *Introductory Plant Biology.* 9th ed. New York: McGraw-Hill, 2003.

Waldman, Carl. *Atlas of the North American Indian.* New York: Checkmark Books, 2000.

Ward, Diane Raines. *Water Wars.* New York: Riverhead Books, 2002

Wells, Neil. *The Atmosphere and Ocean.* W. Sussex, England: John Wiley & Sons, 1997.

Withgott, Jay, and Scott Brennan. *Environment: The Science Behind The Stories.* 2nd ed. California: Pearson Benjamin Cummings, 2007.

Wright, Richard T. *Environmental Science.* 9th ed. New Jersey: Pearson Prentice Hall, 2005.

-------. *Environmental Science.* 10th ed. New Jersey: Pearson Prentice Hall, 2008.

Magazine Articles

Alley, Richard B. "Abrupt Climate Change." *Scientific American,* November 2004, p. 62.

Anderson, J.W. "What Follows Kyoto?" *Resources,* Issue 157, Spring 2005, *p. 14.*

Appenzeller, Tim. "The Coal Paradox." *National Geographic,* March 2006, p. 98.

-------. "Big Thaw." *National Geographic,* June 2007, p. 56.

-------. "Tracking the Next Killer Flu." *National Geographic,* October 2005, p. 2.

Appenzeller, Tim et al. "The Heat Is On." *National Geographic,* September 2004, p. 2.

Bourne, Joel K. "Land on the Edge." *National Geographic,* July 2006, p. 60.

-------. "Fall of the Wild." *National Geographic,* May 2006, p. 42.

-------. "Biofuels: Boon or Boondoggle?" *National Geographic,* October 2007, p. 38.

Caldeira, Ken, Atul K. Jain, and Martin I. Hoffert. "Climate Sensitivity Uncertainty and the Need for Energy without CO_2 Emission." *Science,* vol. 299, 28 March2003, p. 2052.

Chadwick, Douglas H. "Urban Elephants." *National Geographic,* October 2005, p. 98.

-------. "Orcas Unmasked." *National Geographic,* April 2005, p. 86.

Chambers, Jeffrey Q., et al. "Hurricane Katrina's Carbon Footprint on U.S. Gulf Coast Forests." *Science,* Vol. 318, 16 November 2007, p. 1107.

Collins, William, et al. "The Physical Science Behind Climate Change." *Scientific American,* August 2007, p. 64.

Cox-Foster, Diana, et al. "A Metagenomic Survey of Microbes in Honey Bee Colony Collapse Disorder." *Science,* vol. 318, 12 October 2007, p. 283.

Doney, Scott C. "The Dangers of Ocean Acidification." *Scientific American,* March 2006, p. 58.

Doyle, Rodger. "Melting at the Top." *Scientific American,* February 2005, p. 31.

-------. "Oil Haves and Have-Nots." *Scientific American,* September 2004, p. 31.

Duncan, David Ewing. "Pollution Within." *National Geographic,* October 2006, p. 116.

Ehrlich, Gretel. "The Last Days of the Ice Hunters." *National Geographic,* January 2006, p. 78.

Fay, J. Michael. "Ivory Wars." *National Geographic,* March 2007, p. 34.

Finkel, Michael. "Raging Malaria." *National Geographic,* July 2007, p. 32.

Fischetti, Mark. "The Nuclear Threat." *Scientific American,* November 2007, p. 74.

Fishman, Ted C. "Cars That Guzzle Grass." *The New York Times Magazine,* 25 September 2005, p. 86.

Gertner, Jon. "The Future Is Drying Up." *The New York Times Magazine,* 21 October 2007, p. 68.

Gibson, Daniel. "Confronting Eternity." *Calypso Log,* vol. 18. October 1991, p. 14.

Goldberg, Alan M., and Thomas Hartung. "Protecting More Than Animals." *Scientific American,* January 2006, p. 84.

Gugliotta, Guy. "Maya Mysteries." *National Geographic,* August 2007, p. 68.

Halpern, Michael. "Science at the FDA." *Catalyst, The Magazine of the Union of Concerned Scientists,* vol. 5, fall 2006, p. 6.

Hansen, James. "Defusing the Global Warming Time Bomb." *Scientific American,* March 2004, p. 68.

Hawkins, David G., Daniel A. Lashof, and Robert H. Williams. "What to Do about Coal." *Scientific American,* September 2006, p. 68.

Hayden, Thomas. "Super Storms." *National Geographic,* August 2006, p. 66.

Hoegh-Guldberg, O., et al. "Coral Reefs Under Rapid Climate Change and Ocean Acidification." *Science,* vol. 318, 14 December 2007, p. 1737.

Hoffert, Martin I., et al. "Advanced Technology Paths to Global Climate Stability: Energy for a Greenhouse Planet." *Science,* vol. 298, 1 November 2002, p. 981.

Holden, Constance. "Lean Times for Lake Superior." *Science,* vol. 318, 9 November 2007, p. 893.

Holland, Jennifer S. "The Acid Threat." *National Geographic,* November 2007, p. 110.

Kaiser, Jocelyn. "Lawmakers Worry that Lab Expansion Poses Risk." *Science,* vol. 318, 12 October 2007, p. 182.

Kasting, James F. "When Methane Made Climate." *Scientific American,* July 2004, p. 78.

Kerr, Richard A. "Is Battered Arctic Sea Ice Down for the Count?" *Science,* vol. 318, 5 October 2007, p. 33.

Kintisch, Eli. "Tougher Ozone Accord Also Addresses Global Warming." *Science,* vol. 317, 27 September 2007, p. 1843.

Krauss, Lawrence M. "Questions That Plague Physics." *Scientific American,* August 2004, p. 82.

Lyman, Edwin, and Lisbeth Gronlund. "How to Help Terrorists Get the Bomb." *Catalyst, The Magazine of the Union of Concerned Scientists,* vol. 5, spring, 2006, p. 8.

MacKenzie, Don. "Ethanol." *Catalyst, The Magazine of the Union of Concerned Scientists,* vol. 5, fall, 2006, p. 18.

Mallin, Michael A. "Wading in Waste." *Scientific American,* June 2006, p. 52.

Mandelbaum, Robb. "Life after Oil." *Discover,* August, 2006, p. 55.

McKibben, Bill. "Carbon's New Math." *National Geographic,* October, 2007, p. 33.

Michaels, David. "Doubt in Their Products." *Scientific American,* June 2005, p. 96.

Mitchell, John G. "Tapping the Rockies." *National Geographic,* July 2005, p. 92.

Montaigne, Fen, Kennedy Warne, and Chris Carroll. "The Global Fish Crisis." *National Geographic,* April 2007, p. 32.

Myers, M.D., et al. "USGS Goals for the Coming Decade." *Science,* vol. 318, 12 October 2007, p. 200.

Nicklen, Paul. "Vanishing Sea Ice." *National Geographic,* June 2007, p. 32.

Parfit, Michael. "Powering the Future." *National Geographic,* August 2005, p. 2.

Pollan, Michael. "Power Steer." *The New York Times Magazine,* 31 March 2002, p. 44.

Rensberger, Boyce. "Science Abuse." *Scientific American,* October 2005, p. 106.

Robinson, Emily. "Exxon Exposed." *Catalyst, The Magazine of the Union of Concerned Scientists,* vol. 6, spring, 2007, p. 2.

Rogers, Erin, and Spencer Quong. "Leading the Way in Clean Vehicle Design." *Catalyst, The Magazine of the Union of Concerned Scientists,* vol. 6, spring 2007, p. 8.

Sachs, Jeffrey D. "Climate Change Refugees." *Scientific American,* June 2007, p. 43.

-------. "The New Geopolitics." *Scientific American,* June 2006, p. 30.

-------. "The Challenge of Sustainable Water." *Scientific American,* December 2006, p. 48.

-------. "Climate Change and the Law." *Scientific American,* November 2007, p. 38.

Sapolsky, Robert. "Sick of Poverty." *Scientific American,* December 2005, p. 92.

Satyapal, Sunita, John Petrovic, and George Thomas. "Gassing Up with Hydrogen." *Scientific American,* April 2007, p. 81.

Sills, Jennifer. "The Fire Retardant Dilemma." *Science,* vol. 318, 12 October 2007, p. 194.

Smith, Heather L. "Washington View." *The Reporter, Population Connection,* vol. 37. fall 2005, p. 12.

Socolow, Robert H. "Can We Bury Global Warming?" *Scientific American,* July 2005, p. 49.

Tren Kevin E. "Warmer Oceans, Stronger Hurricanes." *Scientific American,* July 2007, p. 45.

Wallace, Scott. "Farming the Amazon." *National Geographic,* January 2007, p. 40.

Ward, Peter. "Impact from the Deep." *Scientific American,* October 2006, p. 64.

Williams, Florence. "Toxic Breast Milk?" *The New York Times Magazine,* 1 January 2005, p. 21.

Yermiyahu, U., et al. "Rethinking Desalinated Water Quality and Agriculture." *Science,* vol. 318, 9 November 2007, p. 920.

Zweibel, Ken, James Mason, and Vasilis Fthenakis. "A Solar Grand Plan." *Scientific American,* January 2008, p. 64.

European Space Agency. "Hole Shrinkage." *Scientific American,* December 2007, p. 37.

"Problems Government Can't Ignore." *Catalyst, The Magazine of the Union of Concerned Scientists,* vol. 6, spring, 2007, p. 15.

"PopPourri." *The Reporter, Population Connection,* vol. 38, summer, 2006, p. 4.

Newspaper Articles

Arenson, Karen W. "Can't Complete High School? Just Go Right Along to College." *The New York Times,* 30 May 2006, Sec. A, p. 1, col. 3.

Baltimore, David. "With Science, U.S. Is Getting Left in the Dust." *The Los Angeles Times,* 29 November 2004, Sec. B, p. 9.

Barboza, David. "Waiter, There's a Celebrity in My Shark Fin Soup." *The New York Times,* 13 April 2006, Sec. 4, p. 3, col. 1.

Barringer, Felicity, and Micheline Maynard. "U.S. Court Voids Fuel Standards For Some Trucks." *The New York Times,* 15 November 2007, Sec. A, p. 1, col. 6.

Barrionuevo, Alexei. "Bees Vanish; Scientists Race for Reasons." *The New York Times,* 24 April 2007, Sec. F, p. 1, col. 3.

Broad, William J. "U.S. Is Losing Its Dominance in the Sciences." *The New York Times,* 3 May 2004, Sec. A, p. 1.

Calhoun, John B. "Population Density and Social Pathology." *California Medicine, The Western Journal of Medicine,* vol. 113(5), November 1970, p. 54.

Dillon, Sam. "Test Shows Drop in Science Achievement for 12th Graders." *The New York Times,* 24 May 2005.

-------. "Literacy Falls for Graduates from College, Testing Finds." *The New York Times,* 16 December 2005, Sec. A, p.34, col. 6.

Friedman, Thomas L. "Losing Our Edge?" *The New York Times,* 22 April 2004, Sec. A, p. 27, col. 5.

Harden, Blaine, and Douglas Jehl. "Ranchers Bristle as Gas Wells Loom on the Range." *The New York Times,* 29 December 2002, Sec. A, p. 1, col. 1.

Johnson, Kirk, and Dean E. Murphy. "Drought Settles In, Lake Shrinks and West's Worries Grow." *The New York Times,* 2 May 2004, Sec. A, p. 1, col. 2-4.

Kahn, Joseph. "Cheney Promotes Increasing Supply as Energy Policy."

The New York Times, 1 May 2001, Sec. A, p. 1, col. 6.

-------. "U.S. Scientists See Big Power Savings from Conservation." *The New York Times,* 6 May 2001, Sec. 1, p. 1, col. 6.

Kahn, Joseph, and JimYardley. "As China Roars, Pollution Reaches Deadly Extremes." *The New York Times,* 26 August 2007, Sec. A, p. 1, col. 1-4.

Karuss, Clifford, et al. "As Polar Ice Turns to Water, Dreams of Treasure Abound." *The New York Times,* 10 October 2005, Sec. A, p. 1, col. 1.

Kristof, Nicholas D. "Chinese Medicine for American Schools." *The New York Times,* 27 June 2006, Sec. A, p. 17, col. 1.

Lueck, Thomas J. "Bloomberg Draws a 25-Year Blueprint for a Greener City." *The New York Times,* 23 April 2007, Sec. B, p. 1, col. 2-5.

Revkin, Andrew C. "No Escape: Thaw Gains Momentum." *The New York Times,* 25October 2005, Sec. F, p. 1, col. 1-4.

Ricks, Delthia. "Air Pollution Found to Damage Arteries." *Newsday,* 21 December 2005, Sec. A, p. 6, col. 1-3.

Robbins, Jim. "Hunting Habits of Yellowstone Wolves Change Ecological Balance in Park." *The New York Times,* 18 October 2005, Sec. F, p. 3.

Schemo, Diana Jean. "Most Students in Big Cities Lag Badly in Basic Science." *The New York Times,* 16 November 2006, Sec. A, p. 22, col. 5

Timmons, Heather. "Britain Warns of High Costs of Global Warming." *The New York Times,* 31 October 2006, Sec. A, p. 8, col. 4.

"A Thrifty Spin in a 99 M.P.G. Car." *The New York Times,* 16 September 2001, Sec. 12, p. 1.

"Literacy Declines among College Grads." *On Campus, The American Federation of Teachers,* vol. 25, March/April 2006, p. 5.

"Students Rarely Ready for College." *On Campus, The American Federation of Teachers,* December 2004/January 2005, p. 6.

Websites

Sierra Club, San Francisco, CA www.sierraclub.org

Carrying Capacity Network, Washington, D.C. www.carryingcapacity.org

Center for Disease Control and Prevention, Atlanta, GA www.cdc.gov

United States Census Bureau, Washington, D.C. www.census.gov

Cable News Network, Atlanta, GA www.cnn.com

Discovery Communications, Silver Spring, MD www.dsc.discovery.com

California Energy Commission www.energy.ca.gov

USDA Forest Service www.fs.fed.us

U.S. Fish and Wildlife Service www.fws.gov

San Diego State University www.geology.sdsu.edu

Intergovernmental Panel On Climate Change www.ipcc.ch

Maine Yankee Nuclear Power Plant www.maineyankee.com

Newsday, Melville, NY www.newsday.com

British Broadcasting Corporation, London, UK www.bbc.co.uk

National Wildlife Federation, Reston, VA www.nwf.org

Negative Population Growth, Alexandria, VA www.npg.org

U.S. Nuclear Regulatory Commission, Rockville, MD www.nrc.gov

Public Broadcasting Service, Arlington, VA www.pbs.org

Sandia National Laboratories, Albuquerque, NM www.sandia.gov

U.S. Dept. of Health & Human Services, Washington, D.C. www.surgeongeneral.gov

Society of Petroleum Engineers, Houston, TX www.spe.org

Union of Concerned Scientists, Cambridge, MA www.ucsusa.org

U.S. Geological Survey, Reston, VA www.usgs.gov

United Nations, NY, NY www.un.org

The White House, Washington, DC www.whitehouse.gov

World Health Organization, Geneva, Switzerland www.who.int

World Oil, Houston, TX www.worldoil.com

Worldwatch Institute, Washington, D.C. www.worldwatch.org

World Wildlife Fund, Washington, D.C. www.worldwildlife.org

World Wildlife Fund, UK www.wwf.org.uk

Yellowstone National Park, U.S. www.yellowstone.com

National Park Service, Washington, D.C. www.nps.gov

Pamphlets

Population Reference Bureau. Washington D.C. World Population Data Sheet, 2007.

Union of Concerned Scientists. Cambridge, MA. IPCC Highlights Series. Climate Change Science, 2007.

Union of Concerned Scientists. Cambridge, MA. Confronting Climate Change in the Northeast, 2007.

Printed in the United States
By Bookmasters